Goethe's Science of Living Form

⌐ The Artistic Stages ⌐

Goethe's Science of Living Form

⁓ *The Artistic Stages* ⁓

Nigel Hoffmann

With a foreword by Craig Holdrege

Illustrated by Pamela Dalton

Adonis Press

Third edition, 2022

Published by Adonis Press
321 Rodman Road
Hillsdale, NY 12529
www.adonispress.org

Adonis Press is a branch of the Hawthorne Valley Association

This publication was made possible through the generous support of the Waldorf Curriculum Fund and the Foundation for Rudolf Steiner Books.

ISBN 0-932776-35-3
ISBN 978-0-932776-35-8

Cover design by Dale Hushbeck
Black and white illustrations by Pamela Dalton
Color Plates by Nigel Hoffmann
Foreword ©2006 by Craig Holdrege

The cover image of *Sisymbrium officinale,* also used in Figure 2 on page 39, is from *New Eyes for Plants* by M. Colquhoun and A. Ewald and is used with kind permission of Hawthorn Press, Stroud, UK, 1996, I-869890-85-X; www.hawthornpress.com

I dedicate this book to those, around the world, who are developing Goethe's way of science with enthusiasm and devotion — in particular, to Dr. Margaret Colquhoun, who was an inspiration for me along this path.

Contents

Foreword

Having just celebrated, on August 28, 1786, his thirty-seventh birthday with friends in the spa city of Carlsbad, Germany, Goethe stole away in the middle of the night, incognito on a postal coach. His goal was Italy. Goethe had already gained fame as a writer and poet. He had served for ten years as a minister in the Dukedom of Weimar. And he had carried out an array of scientific studies. But he needed a change; he felt stifled. His answer was to gain fresh experiences and let the world breathe new life into him.

He crossed over the Alps and arrived in northern Italy, where he wrote in his journal:

> I console myself with the thought that, in our statistically minded times, all this has probably already been printed in books which one can consult if need arise. At present I am preoccupied with sense-impressions to which no book or picture can do justice. The truth is that, in putting my powers of observation to the test, I have found a new interest in life. How far will my scientific and general knowledge take me: Can I learn to look at things with clear, fresh eyes? How much can I take in at a single glance? Can the grooves of old mental habits be effaced? This is what I am trying to discover." (*Italian Journey*, September 11, 1786, p. 21)

These sentences characterize beautifully Goethe's approach to science. First, he had a keen interest in all sensory phenomena. He was a born observer. He wasn't satisfied with single observations, but wanted to get to know things in all their variations. So, for example, when he was traveling — by horse-drawn coach — over the Alps he noticed how familiar species of plants changed in their growth habits: "in the low-lying regions, branches and stems were strong and fleshy and leaves broad, but up here in the mountains, branches and stems became more delicate, buds were spaced at wider intervals and the leaves were lanceolate in shape" (*Italian Journey*, September 8, 1786, p. 15).

But Goethe was not only sensitive to sense impressions. He noticed how he interacted with the world as a perceiving, thinking, and feeling human being. When he asks, "How can the grooves of old mental habits be effaced?" he is aware that how we think about things can hinder experiencing them in fresh new ways. Goethe's scientific writings are full of comments about the relation between the observer and the observed. This sensitivity toward the qualities of the phenomena and to his own interaction with them is captured well in his expression "delicate empiricism" (which Goethe first used in 1829; see Miller, 1995, p. 303). His goal was to let the phenomena speak. To this end he knew he had to be most delicate in the way he applied concepts — so that pre-formed mental grooves did not force the phenomena into particular conceptual frameworks.

Already in his time, Goethe felt that the phenomena of nature were, on the whole, being molded to fit either into mechanistic or teleological frameworks. Goethe certainly recognized the value of anatomy — the dissection of an organism into parts. But he also realized that if you try to build up a picture of an organism starting with the already dissected parts, you end up with a mechanistic picture — the organism as a machine in which the additive functions of the parts "explain" the whole. Such an approach provides only a shadowy image of the reality of a living organism. At the same time, Goethe was unsatisfied with teleological explanations of organisms, explanations that project a divine goal or purpose into things. Such explanations always presuppose an unknowable "beyond" and, like mechanistic schemes, leave essential features of living organisms untouched.

Goethe's desire was "to understand living formations as such, to grasp their outwardly visible, tangible parts in context, to see these parts as an indication of what lies within and in this way to get a hold of and behold the whole" (1817, p. 47). Since everything in the organic world is in a state of flux — developing, changing form, reproducing, aging, dying — we need to become mobile thinkers to gain understanding of the organic world. As Goethe put it, "if we want to reach a living understanding of nature, we must follow her lead and become as mobile and flexible as nature herself" (1817, p. 48). For Goethe, doing science well meant that the scientist must transform his or her own way of knowing to be adequate to the phenomena in question.

In this book, Nigel Hoffmann encourages us to look at nature with fresh eyes and to gain a new kind of mobility in our thoughts. His approach is to lead the reader into different ways of cognizing natural phenomena. He takes as his guide the ancient idea of the four elements — earth, water, air, and fire. Most of us think of these four elements as substances. Hoffmann shows, however, that already the Greeks saw them as ways of knowing — that by looking at the world in terms of water, the water qualities of the world reveal themselves. So Hoffmann characterizes the elements as different modes of cognition, each of which opens up new qualities of the world. If we only look at things in terms of "earth" qualities — such as solidity, enclosed form, inertia — we will miss "water" qualities of flow and ongoing transformation that also inform the world.

So by showing that there are different ways of knowing and choosing modes of cognition that are qualitative, Hoffmann's approach does a manifold service. First, it helps us gain a mobility of thought by moving from one perspective to another. This helps "efface mental grooves." Second, we become more sensitive to the limitations of any one perspective we or someone else might take. Third, by taking multiple perspectives, we can begin to see and appreciate new aspects of the phenomena that might have otherwise gone unnoticed. Fourth, because the four elements are qualities, they lead us more deeply into the qualitative features of nature, which have long been considered off-limits to scientific inquiry.

Hoffmann does not stay with a philosophical and methodological elucidation of a new approach. He has worked at practicing it. The last chapter of the book is dedicated to a landscape study — of the Yabby Ponds in Australia — in which Hoffmann applies the four elements as perspectives to gain an understanding of this unique place. Through his descriptions, the reader can get a sense of how one can actually look at topography, geology, plants, animals, and a whole landscape from four different vantage points, each of which opens up a new facet of the landscape.

Clearly, our human interactions with nature today are often destructive, and yet we yearn for a deeper connection with nature. There is both a disconnect and a desire to overcome it. But, there is a problem. And, to paraphrase Einstein, we can't solve problems by using the same kind of thinking we used when we created them. Since most of our ecological problems stem from thinking of nature as a collection of "things" (resources) outside ourselves that we can exploit at will, we need to transcend that mindset.

Hoffmann's book shows that we don't have to remain caught in traditional habits of thought. We can work to become more attentive to the qualities of nature. In as far as we recognize and take such qualities into our experience, nature as "mere object" disappears. Meeting nature qualitatively means meeting beings and relations that have their own integrity and that warrant our recognition and respect. This is a new kind of scientific-artistic practice. I put my hopes for the future in such practice because it plants seeds of a life-attuned thinking into the world that can help us to act in more life-engendering ways. It is not the "same kind of thinking" that created our present-day ecological problems. I hope this book finds readers who put it to the test, take it into their lives and into their fields of work. We need to develop new pathways into the world.

— Craig Holdrege, October, 2006

References

Goethe, J.W. (1817). "Zur Morphologie: Die Absicht wird eingeleitet." In *Schriften zur Naturwissenschaft*. Stuttgart: Reclam, 1982. (English translations by Craig Holdrege)

Goethe, J.W. (1982). *Italian Journey 1786-1787* (translated by W.H. Auden & Elizabeth Mayer). San Francisco: North Point Press.

Miller, Douglas (ed). (1995). *Goethe: Scientific Studies* (Collected Works, Vol. 12). Princeton, NJ: Princeton University Press.

PART I

TOWARD AN AUTHENTIC SCIENCE

OF LIVING FORM

" ... to my mind there is a great field of science which is as yet quite closed to us. I refer to the science which proceeds in terms of life and is established on data of living experience and of sure intuition. ... Our science is a science of the dead world. Even biology never considers life, but only mechanistic functioning and apparatus of life... "

— D. H. Lawrence[1]

"We study only the corpse in science today."

— Rudolf Steiner[2]

"Everything which man undertakes to perform, whether it is accomplished by words or deeds or otherwise, must spring from all his united powers; everything fragmentary is blameworthy."

— Goethe[3]

CHAPTER 1

THE QUESTION OF METHOD

Art and the Emergence of an Authentic Organic Science

The novelist D. H. Lawrence had the temerity to cast an artistic glance over the life sciences and conclude that they never deal with life at all, only with a dead world of mechanical functions and operations. Views of this sort tend to be regarded by scientists as sentimental or dangerously ignorant — and dismissed outright. Actually Lawrence is not quite accurate because he also asserted that the possibility of an authentic science of the living world is, as yet, entirely closed to us. In fact, long before his time the seeds of such a science had been planted in European culture, but their growth and development had taken place in relative obscurity, mostly beneath the surface of cultural life. This science has now reached the point where its unique character and the wide-reaching significance of its aims can be far more easily discerned.

Organicism — as a philosophy or way of thinking — has been a powerful force in the evolution of Western civilization. A coherent stream of such thinking can be traced at least as far back as the natural philosophy of Aristotle and has amongst its more recent representatives the philosophers Spinoza, Bruno, Hegel and Schelling, and the poet and natural scientist J. W. von Goethe. Goethe, around a century before Lawrence, had come to realize the inability of rationalistic science — the science of the eighteenth century Enlightenment — to come to grips with organic nature. Nourished by the German cultural movement known as *Naturphilosophie*, which brought together scientists, artists and philosophers in highly fruitful relationships, Goethe reached toward an artistic form of science, a science that is adequate to the world of the living, a science in which *the human artistic faculties are formed into organs of cognition.*

It would be simply erroneous and misguided to claim that a true life science exists just because scientists occupy themselves with the analysis of organisms — plants, animals and humans — or because science is now acquiring a sophisticated knowledge of the structure and functioning of genes. Of course science provides comprehensive definitions of what functions distinguish one living thing from another, but these always turn out to be machinelike functions or mechanisms — forms and processes which can be isolated, analyzed, computed and manipulated, in precisely the same way as machines are analyzed and engineered. This science views only "the apparatus of life," as Lawrence puts it. In this book I will endeavor to show that a science of the living realm calls for art, that it must be infused with the artistic — and this means something far more than artistic practice as a kind of adjunct to the otherwise totally inartistic procedures of science. It calls upon human artistry and scientific discipline to reawaken to each other in such a way that the whole human being engages in the act of cognition.

In the contemporary world, the cognitive function of art is generally overlooked, and one runs the risk of being branded naïve even to suggest that art has got something fundamentally to do with the practice of science. Besides, the lifestyle of the analytical scientist seems as remote as can possibly be from that of the artist — say, the painter in the pigment-strewn studio, to take an example from popular imagery. From the side of art, many influential artists and art theorists of the twentieth century have argued, even insisted, that art does not and never will fit in with society's practical interests of understanding and developing the world. In the course of European cultural evolution a rift opened between the arts and sciences which gradually broadened into an abyss.

It is nevertheless true that, at the dawn of the new millennium, art's position is shifting. There is still the respectable and important role of art as entertainment and aesthetic upliftment, and there is still a place for the traditional religious role of art as revelation. Art plays many other roles in society — in illustration, design and so on. Beyond these, art still lives a wild and free existence in the fringe realms of the counter-culture, the avant garde or whatever name is given to that impulse in art which feels itself to be *the* leading edge of the present into the future. But there is another way forward for art, art in an emergent mode of its being that is harder to recognize because it is only now becoming a defined way of working. This is the way that knows of no more important aim than to serve an authentic understanding of the living world. One writer puts it like this: "Art . . . sacrifices its independent existence for the service of the cognitive task of science."[4]

Initially art can play a vital role in helping science to come to a clearer perception of itself.[5] To the extent that the philosophy of science works only in a theoretical way as it strives to come to grips with the nature of science — to deconstruct it through critical analysis and discourse — what appears as radical and liberated is, from another point of view, only a perpetuation of the same: more theory, different theory, but still just theory. The theoretical mode *itself* needs to be perceived, its whole gesture and mood must be apprehended, and this requires a different organ of understanding from that of theoretical consciousness. This is why an artist like Lawrence, with his highly individualized artistic sensibility, could look at science and grasp so readily what theoretical thinking cannot do out of itself. Lawrence could perceive, in its pervasive tone and atmosphere, in the whole manner of its striving, that theoretical science cannot "see" life and therefore cannot be responsible *for* life.

Lawrence's observations are insightful and provocative yet go no way toward showing how a true science of the living can come into being; that was not his mission. Turning to the lifework of Goethe, however, we have before us the work of an artist who not only recognized the deadening character of theoretical science but actually created a new form of science. Goethe lived at the time when the notion coming to the forefront of the scientific worldview was that the human being, indeed the whole organic world, is a complex machine.[6] A lot of attention is now being paid to Goethe's theories of color and morphology and his views on the history of science, but his central and great deed was to show that an "exact sensory imagination" — which "art is unthinkable without" as he wrote — can become an organ of scientific consciousness.[7] It is because Goethe was able to achieve an unprecedented awakening of artistic consciousness within the domain of science that the philosopher and artist Rudolf Steiner was moved to describe him as "the Copernicus and Kepler of the organic world."[8]

Most scientists today would still declare that life itself, in as far as it exists beyond bio-chemistry, cannot be thought, that the actual "livingness" of an organism is a matter of values, feelings or vague intuitions and therefore a meaningless notion as far as scientific understanding is concerned. What this points to is the fact that science in its "classical" form cannot think life. A scientific thinking that is mechanical and logical perceives only that dimension of nature which is mechanical and logical; thinking and being correspond to each other in this way. Such a thinking gradually emerged in the cultures of the ancient world, gathered force in Greek civilization and evolved over a long period to reach a climax around the time of the Renaissance in Europe, after which it became the foundation of modern science.

But there is also a thinking that belongs to the realm of the living and proceeds in terms of life, a genuine organic thinking. This thinking has also evolved over a long period but in a more or less concealed fashion during the time when rational scientific consciousness was in the ascendant. We may consider, for example, the dynamic, pictorial kind of thinking required to understand music. As will be seen in what follows, musical thinking is the precursor of an exact method of perceiving the dynamic structures of living form. Throughout European history a heightened pictorial-musical thinking has been cultivated within cultural streams of a somewhat esoteric variety where the artistic and scientific have always been closely intertwined. In his youth, Goethe himself took a deep interest in alchemy and there can be no doubt that at least some of the ideas he worked with in his studies of natural forms were derived from this source of inspiration.[9] Later he immersed himself in the exoteric streams of organic thought. Aristotle had originally talked about an indwelling life principle which he called the *entelechy* — but the question has been ever since: how could such a principle be perceived, how could it be proven to exist? Goethe's investigations of nature represent a decisive development in the history of the science of living form because of the way he was able to distill out of the artistic and scientific currents of European culture a mobile imaginative form of thinking adequate to living form and provide it with a sound methodological footing in the context of rationalistic science.

More recently, the work of Rudolf Steiner has had considerable significance in the development of an authentic life science; Steiner had a profound appreciation of what is artistic in Goethe's way of science and was able to develop this in his own work to a high degree.[10] It was Steiner who warned that "the schism between science and art has disrupted the very being of man."[11] As indicated by Steiner, there are four discrete stages in the cultivation of an authentic living thinking.[12] Conventional scientific intelligence he terms *physical thinking*, the thinking that develops from an apprehension of cause and effect processes in nature and the laws of nature as they are ordinarily understood. Steiner calls *Imagination* the transformation of the human *sculptural* sensibility into an organ for understanding the plastic nature of organic form, how living things are molded into existence. Imagination already moves far beyond the domain of conventional scientific rationality. *Inspiration* is what Steiner calls the metamorphosis of the *musical* sensibility into a faculty for understanding the *gestural* structure of organisms. Lastly, he speaks of the transformation of the *poetic* sense into the organ of cognition he calls *Intuition*. Through Intuition the formative idea in living form comes to light. As we shall see, the practice of the different arts — in particular modeling, music, poetry and the art of speech — has a vital role to play in Goethean phenomenology as developed by Steiner.

Over the last hundred years many writers have lamented that the arts have become wayward and have gone about devaluing what tradition deemed true, beautiful and good. The twentieth century saw everything, from the bizarre, to the self-negating, to the violent and pornographic, raised up in the name of art. The "problem" lies in the relentless striving for freedom; it seems the human artistic spirit simply cannot rest content, cannot feel true and whole, unless it is in touch with the pure fire of creative freedom. Yet only through freedom can there be individual moral choice, and only by virtue of an achieved independence is it possible to willingly relinquish this independence as a deed of conscience. The meaning of art sacrificing its independent existence in order to serve the cognitive task of science has nothing to do with obligation or some kind of imposition on the creative freedom of the artist. It springs from the recognition that a true science of living form must have art.

Goethe and the Phenomenological Method

The gulf between science and art shows nowhere more clearly than with respect to *method*; it is here that the greatest obstacles appear before those who would negotiate that divide. The problem can be put in a nutshell: how does one unite that which is essentially free with that which is essentially determined? Classical scientific method is understood to be determined by the concrete form of nature itself — that is the basic rationale without which it is nothing. This is the correspondence theory of knowledge, the idea that the application of scientific logic leads to knowledge because it accords with the way nature itself works. Method in science means the intellectual process which eliminates all caprice and allows for a rigorous alignment with what is actually objectively the case. This is not to say that the scientist is incapable of art or creativity — let us call it "inspiration" — only that inspiration is "extra-scientific," unstructured, outside the paradigm of the scientific method.[13]

Now the opposite principle obtains for art. Except in the case of representational art, which verges upon scientific illustration, art is nothing unless it allows for the exercise of creative imagination and inspiration through its methods, its techniques. Here imagination is entirely "free" and resists being strictly governed by "what is objectively the case." The *knower* and the *creator* — these are the two polar aspects of human nature which are indicated when we speak of science and art. One side of the polarity points to the world of solid, objective factuality, the other to the free flight of the imagination. Some would say it is consoling and even fashionable to talk about the unity of art and science but not so easy to conceive of a *method* that unites creative freedom and objective realism. Such a unitary method of working is precisely that which Goethe strove to bring into being through his investigations of living things.

In his essay *Empirical Observation and Science* Goethe presents a sketch of a method to which he says he has made an effort to remain true.[14] He begins with the "empirical phenomenon … which everyone finds in nature." The scientist raises the "empirical phenomenon" to the "scientific phenomenon" through observing the phenomenon under a variety of conditions and through experimentation. Finally the "pure phenomenon" emerges, not as something that can be isolated or abstracted from the phenomenon, but as that which illumines the phenomenon which was hitherto opaque to unscientific perception. In this particular essay we are not given a lot of information about what this "pure phenomenon" might be (in other places he refers to it as the "archetypal phenomenon"). Goethe speaks of the "self-distilling" process of common human understanding "as it ventures to apply itself to a higher sphere." Plainly this method involves something other than a deductive or inductive reasoning process; the implication is that thinking itself undergoes some kind of purification, a qualitative transformation, such that it is able to perceive the phenomenon in a "higher" dimension of itself.

Elsewhere Goethe classifies four types of people who ask different kinds of questions concerning the natural world.[15] Firstly he describes the "utilitarians," those who seek a knowledge of nature out of a sense that everything can serve human purposes. Then there are those seeking knowledge for its own sake, working in predefined fields like the utilitarians yet without such pragmatic intentions. These are characterized by their "quiet, objective gaze, restless curiosity, and clear understanding." The third kind of people he describes are the "intuitively perceptive," whom he considers as having reached a creative stage. These "seekers of knowledge," he observes, tend to deny the function of the imagination — yet "before they know it, they have to call on the imagination's creative power for help." Finally, there are "the comprehensive (die Umfassenden)," whom he calls "true creators"; these are "productive in the highest degree." With regard to the work of "the comprehensive," he speaks of the "productive power of the imagination combined with all possible reality."

It is this last statement which appears most radical and, at first glance, most perplexing. For here we see combined in almost casual fashion principles that seem fundamentally at odds with each other — the methodological study of objective reality and the creative, unrestricted play of the imagination. It is important to note that Goethe is not talking about the "extra-scientific" role of creative imagination and inspiration — for example, in the production of new hypotheses. Rather, he is suggesting that objective analytical science *advances to a creative or imaginative stage.*

If we take this progression at face value, at least for the time being, a definite process is apparent which can be aligned with Goethe's statements concerning the distillation of the "empirical" into the "pure phenomenon." We might call this the movement from the determined to the determining, the constituted to the constituting, from the objectively structured to the creatively free. He even goes so far as to assert that the "comprehensive" come to a perception of the "unity of the whole" into which nature then must fit itself![16] If ever he made a statement that will alarm the scientific realists it is this one. Artists, however, will understand this very well, for it is the way of all artistic formation. The materials of nature are gathered and remolded in accordance with the creative imagination.

With these descriptions of the scientific process in mind, we can now turn to a more detailed account of Goethe's experimental procedure. In *The Experiment as Mediator between Object and Subject,* he describes "the method which enables us to work most effectively and surely."[17] First he talks about starting with empirical evidence, which is gathered in an exacting, even scrupulous fashion — this we could say is the transition from the "empirical" to the "scientific" phenomenon. The task of the scientific observer is then "to follow every single experiment through its variations." Each observed fact is found to have a connection to another; this leads to a manifolding of the original evidence through a series of contiguous experiments deriving from one another. Evidence is not isolated, but seen comprehensively in a whole context. He then says: "Once sequential evidence of a higher sort is assembled . . . our intellect, imagination and wit can work upon it as they will; no harm will be done . . ." Here we have the same seeming contradiction. At the highest stage of the experimental process, after having labored methodically to perceive lawful relationships between entities, we reach the point where we can think creatively within the framework of this lawfulness; we achieve creative freedom.

Goethe's method has been called phenomenological although he did not use that expression himself. One twentieth-century philosopher, Fritz Heinemann, writes:

> [Goethe's] method is genuinely phenomenological. It begins
> with phenomena, proceeds through them, and ends with them,
> returning at the last from the Ur-phenomenon [archetypal or
> pure phenomenon] to the particulars whose claims have not at
> any point been abrogated.[18]

One may be drawn to compare Goethe's indications with pheno-
menological methods set forth by many thinkers of the twentieth century,
deriving principally from the work of Edmund Husserl. This comparison
can be useful, especially if it is not embarked on with the preconceived
idea that Husserl's phenomenology represents the standard by which all
other phenomenologies rise or fall. Goethe's stages correspond to what
Husserl called the process of "phenomenological reduction," leading from
the factual, concrete world to the stage where the imagination is able to
grasp the law by which the phenomenon is creatively constituted.[19]

Martin Heidegger has depicted the striving of all phenomenology in
the following way:

> [E]ntities must ... show themselves with the kind of access
> which genuinely belongs to them.[20]

This formulation indicates a definite moral disposition or comportment
that marks the beginning of the phenomenological project. It comes through
clearly in the writings of Goethe when, for example, he declares that "we
cannot find enough points of view nor develop in ourselves enough organs
of perception to avoid killing it [the phenomenon] when we analyze it."[21]
Goethe writes that the true botanist

> must find the measure for what he learns, the data for judgment,
> not in himself but in the sphere of what he observes.[22]

It was a moral attitude he shared with his colleagues in the *Naturphilosophie*
movement. The philosopher Schelling insisted that a phenomenon must
not be

> turned, twisted, narrowed, crippled so as to be explicable, at all
> costs, upon principles that we have once and for all resolved not
> to go beyond.[23]

This attitude could be described as the foundation and *raison d'être* of
Goethean phenomenological science — to find the method which truly
belongs to the nature of the living things we are studying.[24]

Toward an Authentic Method in the Life Sciences

The direction and intent of Goethe's scientific endeavors can now be brought to a sharper focus. We have already seen that Goethe was specifically concerned to find his way toward an authentic life science, an appropriate method by which living beings can be investigated. Goethe was acutely sensitive to the "mechanization of the world-view" in his time and, in particular, saw in the hypothetico-deductive methodology of Newton's science a heinous imposition upon the living being of nature. It was his artistic sensibility that led him to an insight that took possession of him as a scientific obligation and moral responsibility; he saw that an organism is, in its essence or primary nature, *free* and *creative*. Of course, an organism like a plant or animal is not free and creative in the way a human being is. Goethe was referring to what we may call the "wholeness" or "inner completeness" of the organism. Freedom and creativity in organic existence means that an organism comes forth "out of itself" and is not merely the product of something else.[25]

A machine is not an organism and is not free in any sense of the word; it is not free because it is entirely determined by something outside itself. A clock, for example, is designed by an engineer and set in motion by an applied force. Computers are machines which appear to allow for a manner of self-evolution, but in truth they depend on what is programmed into them at the beginning; computers have only a virtual life. An external intelligence — the maker or engineer — translates the idea of the machine into a form whereby the parts work as an interrelated totality when set in motion by an applied force. However, the idea, the formative impulse, is not *in* the machine itself, incarnated, as it were; the machine is therefore not self-creative. The idea is in the mind of the creator and the machine is an image or expression of this idea.

It is not difficult to understand why the notion of mechanism was elevated to the status of cosmological viewpoint in an age when science sought clear, logical explanations for all things. The mechanistic worldview is often associated with Descartes, but, going further back in Western metaphysics, its origins can be found in the neo-Platonic outlook. This world-view holds that material things are mere semblances, products of the generative Idea or archetype which is the reality itself and which exists in a metaphysical world. As with machine production, the Idea remains separate, in the mind of the divine maker (the Demiurge), and is not incarnate in the physical thing. This view has been called a "productionist metaphysics" by more recent thinkers.[26]

After Descartes, through the steady application of analytical techniques to the forms of nature, the "productionist metaphysical" view was seemingly overcome. What actually happened was that science became less speculative, more experimental, and the "productionist" viewpoint became more concealed within scientific methodology. Science sought to eliminate from its sphere of operation anything not immediately verifiable, which means: traceable to a definite cause. What modern biological science has developed is a sophisticated view of nature as machine. Rather than seeking causes in the unverifiable metaphysical world, it sought them in the evolutionary past, and, ultimately, in the Big Bang. In this picture of cause and effect processes that are said to ultimately lead to the existence of organisms, there is no longer any conception of purpose or intelligence at work, only blind mechanical process. What remains the same, however, is the mechanistic outlook or "productionist metaphysics" expressed in this style of thinking. The organism is seen to be caused by something outside itself, whether that "outside" be the metaphysical world, the evolutionary past, or the external environment.

Whatever is deemed to be caused, or produced, becomes intelligible through a logic of cause and effect exemplified by the hypothetico-deductive method made famous by Newton. A phenomenon is said to be explained when a causal relationship is proven. Each aspect of the form and behavior of a living being can be submitted to the analysis of the hypothetico-deductive methodology, and in that way science speaks of slowly but surely "unraveling the mechanisms of nature." Now, it is quite true that this method has undergone considerable refinement since the time of Newton; for example, modern systems theory seeks to take into account the complex interactions of a living field of component parts, through studying linkages and feedback responses. On the face of it, this theory appears as a more organic model because it deals with self-regulation, which is certainly an aspect of living organization. Yet the complex entity which works through internal and external feedback processes is still a machine, and, as the philosopher Henri Bortoft has pointed out, the idea of a "living system" is a contradiction in terms. He shows that systems theory is actually a form of analysis derived from the machine notion of nature and that a *living* organism cannot be so analyzed and explained.[27]

With respect to the new fields of systems, complexity and chaos science, one must proceed with caution because claims are made here for an intimate connection with Goethe's organic way of thinking.[28] To take an example from systems/complexity science: the biologist Brian Goodwin describes a collection of organisms such as a swarm of ants as an "excitable field," in a state of chaos or formlessness. Due to the influence of a "strange attractor" the field goes over, apparently spontaneously, to a state of order expressing a definite form. The form, writes Goodwin, comes from nothing, meaning that its appearance is spontaneous, that it is not *caused* by the long process of natural selection as described by the Darwinian model; thus the word *creativity* is gaining currency in this area of research.[29]

What we have before us in this example is simply a process that has not yet been adequately explained — which is why the "attractor" has been provisionally dubbed "strange." Newton was at a similar point when he was trying to explain the parabolic shape of the movement of projectiles; the "strange attractor" in that case was later called a "gravitational field of force." Between the initial "excitable field," particular environmental conditions, the "strange attractor," and the final ordered state of the organic form, a mechanism is at work which science is now seeking to unravel. There is nothing here which in principle cannot be explained, given time and the development of high-powered computers to deal with the complexity of the processes involved. Indeed, to become a rational science on a par with physics is the ambition of this science.[30] Brian Goodwin makes the aim of his work in the complexity sciences of organic form explicit: to develop a mathematical theory of morphogenesis (organic formation) which will produce "an explanation of the type provided by Newton for planetary motion."[31] Yet it is precisely the Newtonian hypothetico-deductive way of taking hold of living phenomena in order to explain them that Goethe himself so vehemently opposed!

Organisms are highly complex entities and the mechanisms involved are complex; we have every reason to suppose that many of these mechanisms are entirely beyond our ken at the present. Such, indeed, is what keeps analytical science moving forward so vigorously. But it is another matter to suggest that systems or complexity theory has anything essentially to do with what Goethe meant when he declared that he could *perceive* the inner completeness of the organism, its living idea or formative principle. And this is the point where it also becomes necessary to decide whether *art* has anything at all to do with modern "cutting edge" biology, or whether, by even considering art in this connection, we are really only casting a sentimental glance back at "Goethe-the-Romanticist" who did not have the combination of high-powered computers, non-linear mathematical logic and the sophisticated techniques of analytical biology at his disposal. It is not a matter of doubting whether systems and complexity biology have, in some way, been *inspired* by Goethean organicism. What is questionable is whether these sciences actually represent Goethe's way of doing science and, especially, the implication that they have *superseded* Goethe's methods.[32]

What is perhaps surprising is that, when Goethe spoke of intuitively perceiving "the idea" and "inner completeness" of the organism, he meant precisely *that which does not demand explanation*. It would be absurd to go around stating that the world is constantly, and in every respect, calling upon us to seek explanations. This is only one limited aspect of human experience and in many other respects the world does not move us in that way at all. We are also drawn to entirely other forms of revelation, other modes of truth-seeking and truth-saying. By definition, that which has the character of "inner completeness" *is not deficient* and does not stand in need of an account which explains it in terms of something other than itself (as resulting from cause and effect relations). Only the world in its *mechanical* nature moves us to seek explanation, for it is evidently not complete unto itself — it has a cause outside itself which needs to be determined in order that its function or operation can be evaluated (and then its use-value appropriated). Only when this is the case does finding an explanation bring satisfaction.

The human artistic sensibility readily understands this notion of "inner completeness" — it describes the experience of art and accounts for the immense cultural prestige which is attached to this experience. An artwork is produced by the artist without having any operational value; it is not purchased primarily because it is going to be useful to the purchaser or because it can *do* anything. We encounter a work of art and may learn to understand it without feeling in the slightest degree moved to *explain* how it operates. This is because in art truth makes its appearance, "sets itself to work," in a manner entirely different from that of a tool or machine.[33] We may speak of the increasing transparency, the heightened presence or luminosity of a work of art which comes about through the process of interpretation. By contrast, facts and explanations pertaining to a machine accumulate over time as the logical steps of an analytical process "loosen" the structure so as to make the function of every part intelligible.

Aristotle tried to show how the imagination works in the understanding of living things as much as it does in the understanding of art. He argued that artworks and organisms have an indwelling idea, a living formative principle which brings them into being — in this way he linked artistic and natural creation.[34] On the level of universal life, he spoke of this organic principle as the *unmoved mover*; on the level of the particular living thing he called it the *entelechy*, the indwelling or incarnated idea, that which is causing the organism without itself being caused.[35] Since Aristotle's time philosophers of nature have pondered much on this connection of art and organism. Kant is especially notable for the way he sought to show that works of art and organisms demonstrate a character of completeness or wholeness. Kant wrote that the parts of an organism "produce a whole by their own causality" and he too divined that the same principle of coherence works in both organic and artistic form.[36] Later the philosopher Schelling was also to write that "[t]he basic character of organization is that . . . it is at once both cause and effect of itself."[37] The wholeness of an organism and work of art cannot be accounted for by the sum or accumulation of cause and effect linkages. When these philosophers declared that an organism is its own cause, they meant that it "comes forth out of itself," is complete unto itself, is self-creative and "free," that in its essential nature it is not determined by anything other than itself. This view remains valid even though the science of ecology has subsequently shown how the particular form of an organism is related to the particularities of its environment or habitat.[38]

The problem is that Kant and Schelling, as philosophers, were bound to express themselves discursively, in the logical terms of a philosophical argument. But Kant also hinted at a thinking which he himself was not able to carry out, yet which he admitted was a possibility of the human mind; this is a thinking which is not discursive but intuitive. Kant calls it the *intellectus archetypus*, the thinking which understands a living being, not by building up an argument from logical linkages between parts, but by thinking "from the whole to the parts" — that is, creatively, in the manner in which the whole or archetype creates its own parts.[39] Schelling later tried to describe the *intellectus archetypus* by alluding specifically to the creative activity of the artist, art being "the one place where producing and intuiting fully coincide."[40] This intuitive artistic thinking to which these philosophers were pointing is what Goethe simply *did* when he exercised his artistic faculties in the study of the natural world.

Every part of a machine acts on other parts; thus, analytical science has shown that a plant embodies a vast number of separate parts (cells, organs, chemicals etc.) which act on other parts in the most complicated ways. But that does not complete the view of the plant because, from a higher point of view, a plant is a wholeness in which there is nothing separate from anything else. A plant is a "one" or whole in a manner totally different from a machine for the simple reason that it is *alive*. A plant is a living whole, meaning that its members are not merely acting *on* each other (to form a coordinated mechanical system) but are deriving *from* each other, creating each other. Something which is alive grows — that is, it becomes something else from out of its own being even while remaining the same entity. A cactus develops a flower, but it is still precisely the same cactus plant. This is an entirely different phenomenon from that of a machine's functioning which changes when different functional elements are added to it, each addition making it a slightly different kind of machine. Just this observation is enough to confirm that it is *impossible* to have an authentic life science of plants, a true botany, which comprehends organisms only in terms of mechanism. Growth is a fundamental phenomenon of the organic world, and a true life science needs to be able to adequately "think growth."

The phenomenon of growth does not require explanation any more than does the development of a melody. One note is not *acting on* another note, *causing* another note — any one note has musical meaning only in terms of every other note in the piece and this is how the piece evolves. The truth-value of a melody comes to light in the way it makes present, perceptible or understandable, the phenomenon of growth. We gain knowledge of the growth and formation of the flower through analysis of its chemical constituents, through comparative anatomy, through exploring hormonal and genetic processes. But when the process of growth is considered in itself, it is found to be continuous, unbroken, whole, not a composite of cause and effect processes. The roots do not *cause* the stem, the leaves do not *cause* the flower, the flower does not *cause* the fruit. The genetic replication processes in plant cells do not *cause* the plant any more than does the process of photosynthesis — both are functional aspects of a continuous and whole life process.[41] We accumulate knowledge of mechanisms but we only *understand* a living plant out of the wholeness of its form, its total "melody." The plant is in a definite sense creating itself, coming forth out of itself, and this process can only be seen by a thinking which thinks from the whole to the parts, an *intellectus archetypus*.

As already discussed, a machine is made up of functional components related through cause and effect interactions, of parts acting on other parts. The idea of that particular machine, as it originated in the mind of the maker, can be deduced from the functional structure of the machine but is in no sense *in* the machine, forming and maintaining it. The value or meaning of a machine is only functional, and its function always relates to an end result which lies outside the machine. A work of creative art is not a machine because its produced parts or elements do not act *on* one another but subsist within the coherence or wholeness which is the work. In art the idea is the thing itself; the artwork is the idea incarnate — this is its value or meaning. Goethe called the work of art the "spiritual-organic" — something not actually alive but bringing the principle of organic creation to the imaginal/spiritual level, which makes it comprehensible to the human being.[42]

An organism is neither a machine (or complex of mechanisms) nor a work of art in the way we normally understand these things. Machines and artworks are human creations, artifacts, expressions of the human spirit. But it is true to say that, through different comportments of thinking, the organism may reveal itself in the manner of a machine and in the manner of a work of artistic creation. Through analytical thinking it shows itself as a complex of cause and effect relations (mechanisms); through intuitive thinking it shows itself as self-creative process. This does not mean that human thinking merely invents or constructs models, subjectively and arbitrarily, which it then imposes on nature. Rather, the different comportments of thinking and the different dimensions of the living phenomenon belong together, provide the conditions for each other.

Remaining within its analytical mode of thinking, science will fail to grasp how the mechanical is taken up within the organic, how organic, creative laws hold sway and subsume the laws of cause and effect. The metamorphosis from the mechanical to the creative, which is the genesis of living form, is also the metamorphosis which scientific thinking needs to undergo to attain to a living form of thinking. The manner in which the mechanical is subsumed by the creative in living form is easiest to illustrate with regard to the human being: the actions and processes of skeleton, muscle and nerve, as well as every biochemical process at the cellular level, are subsumed by the creative intentions, ideas and actions of the human being which are manifestations of the human "I" (the activities of speaking, thinking, playing or creating, and so on). To rightly understand the human being, our thinking must enter into the way the human "I" determines the form and function of the human organism in its entirety, including the upright stance and the form of the different organs. As Ernst-Michael Kranich puts it: "The "I" as inner force of uprightness takes hold not only of the human skeleton, but also of the entire internal organization."[43]

If we look at animal life we see that there is also a principle that governs and subsumes the mechanical and physical — we call this animal "desire." Through its "body of desires" or sentient life the animal is constantly creating itself or determining its own existence. With plants, we see the creative principle working at a "lower" level: the plant is continually creating itself in the way described above — through coming forth out of itself (growing). In general terms, all form and process in the living world is serving a "higher" organic principle (creativity) in the different modes it assumes in the plant, animal and human worlds. The twentieth-century philosopher of science Michael Polanyi has shown that the physical form and mechanical functioning of plants, animals and humans are subsidiary to and determined by the principles of growth, sentience and personhood respectively.[44]

Thus, strictly speaking, it is incorrect to say that the mechanical is a part or component of the organic realm, and it is certainly not true that living form can be reduced to the mechanical, or that living form can be understood in terms of the mechanical and logical. It is like analyzing harmonic logic and construction and then saying that one can understand music in terms of this logic. For the thinking which participates in the living reality of music and becomes a truly musical thinking, it is clear that the living nature of music itself *creates* or *has created* the logic of harmony, all harmonic construction. As with biological form, knowledge of human biochemistry or physiology serves only as a foundation; to understand the living reality of the human being (of which the "I" is the central aspect) there must be a "turning" in thinking, a movement from the analytic to the creative. For it is life which is creating and determining the biochemistry, the physiology. The human "I" cannot be understood in terms of biochemistry or physiology.

Goethe's phenomenological method now takes on a new meaning: as it moves from the "empirical phenomenon," to the "scientific phenomenon," toward the "pure phenomenon," it discovers a form of investigation which accords with organic nature itself. The "empirical phenomenon" is the everyday knowledge which recognizes the objective, physical reality of organisms and their activities and processes. The scientific work begins with precise, objective descriptions and with the experimental work of ordering the phenomena through different "modes of representation," modes which the phenomena themselves demand.[45] The principle of mechanism is one such mode: it is understood by the aspect of the human mind which is itself mechanical — the intellectual/analytical mind. But the resulting mechanistic theory (or any such logically generated theory) is not the actual life of the organism and therefore does not constitute the reality of the organic.

In Goethean phenomenology we approach living wholeness, the essence or creative idea (*entelechy*) of the organic, through imaginatively "dwelling" within the parts and processes in such a way that we learn to see how the whole is creating the parts and processes. At this stage there must be a "turning," a metamorphosis from the logical into the imaginative/intuitive mode of thought. The principle of creation is understood by virtue of the aspect of the human being which is itself creative. This is the *imagination*, which is able to participate in the creative processes in nature and thereby grasp how the creative subsumes and determines the mechanical. At this "highest stage," understanding means "productive power of the imagination combined with all possible reality."

A Goethean Methodology through the Elemental Modes

The organism is a living whole and our thinking in relation to the organic, our method of investigating living form, must itself be adequate to the nature of the organism. This is the conclusion we arrive at when we declare with Heidegger that the task of an authentic science is to allow entities to "show themselves with the kind of access which genuinely belongs to them." We have seen that Goethe's phenomenological science represents just such a "living" method in the way it unites the intellectual with the imaginal, the scientific with the artistic. *When the whole of the human being becomes active in thinking, then the wholeness of the organism comes to presence.*

Husserl spoke of the mutuality of "what is known" (*noema*) and "how it is known" (*noesis*), and Henri Bortoft has developed this concept in order to make clear what is significant about Goethe's way of science.[46] But what Husserl is bringing forward is no twentieth-century discovery; indeed, the mutual nature of the "how" and the "what" in cognition is one of the great seminal insights of Western culture. Husserl was raising to a new consciousness something which was known in the ancient world as the principle of "like knows like." This is how the fifth century B.C. sage Empedocles put it:

> We see Earth by means of Earth, Water by means of Water, divine Air by means of Air, and destructive Fire by means of Fire.[47]

Aristotle was well aware that Empedocles's statement derived from the ancient principle of "like knows like." He writes: "For it is because the soul had *cognition* of all things that [the pre-Socratics] compose it from all the primary principles [Elements]."[48] Aristotelian science was still largely based on the four Elements even though it marks the transition to the more theoretical form which we now call science. It is very difficult for modern scientific thinking to grasp what Aristotle meant when he spoke of this necessary relationship between the human mind and the Elements.

The Elements are going to take a central place in the following discussion of Goethean methodology but not in any sense out of a desire to seek for "answers" in the pre-scientific past. Empedocles was certainly not a scientist in the way we mean it today, and his questioning concerning the world was something quite different from what we call "scientific reasoning." Yet Empedocles's seemingly naïve and abstruse statement hides a truth which brings light to a fundamental problem of the modern scientific disposition. Science presumes that the totality of world phenomena can be brought within the framework of a single way of thinking. However, the ancient wisdom of "like knows like" recognizes a primary qualitative relationship between the knower and the thing which is being known. The task of a genuine life science is to understand the different gestures or dispositions of knowing which are available to it and to determine the relationship of these ways of knowing to the phenomenon under consideration. This task stands at the *beginning* of any scientific endeavor.

The Elements are a way of understanding and entering into the different "dispositions of thinking" which belong to Goethe's way of science. It was stated above that the principle of mechanism (*noema*) is understood through the aspect of the human being which is itself mechanical (*noesis*) — the intellectual/analytical mind. It was also said that the principle of creation (*noema*) is understood through the aspect of the human being which is itself creative (*noesis*) — the imaginative/intuitive mind. These two represent a polarity: the "created" and the "creating." Aristotle (following Empedocles) said that the human soul (or mind) is whole and that this wholeness is composed of four cognitive dispositions. The thinking or mode of observation we mean when we speak of the intellectual mind is what the ancients meant by Earth. In this sense, Earth sees Earth. The mode of cognition we mean when we speak of the imaginative/intuitive mind is what the ancients meant by Fire. Thus Fire sees Fire. Between Earth and Fire are Water and Air — and, as we shall see, these represent forms of thinking which mediate between the polarities of Earth and Fire.

The human mythic imagination divined that the Elements together constitute the Oneness of Life or "world soul" (*anima mundi*), this being pictured in the form of a sphere, an image of the wholeness of life, "well rounded" and complete unto itself. The human soul or mind was conceived of as being correspondingly rounded (whole) and composed of the Elements.[49] Ancient wisdom understood that the *anima mundi* differentiates into the forms of all things even while remaining one and the same — which is precisely how the organism has been depicted above, as self-creating, growing and differentiating into its different parts even while remaining the same organism. The Elements were understood as divinities or archetypal creative powers (*archai*), the powers that allow the Oneness of Life to differentiate itself into the particular forms of life. It is important to realize that the mundane phenomena of earth, water, air and fire are not the Elements; rather, these are for us the most immediate or concrete *expressions* of the Elemental principles. In the human soul these principles come to expression as what might be called "soul dispositions" or, traditionally, as the temperaments. Thus the ancients had a definite way of thinking about living nature, but this was not a picture derived from a theoretical kind of thinking. Rather — as many commentators have observed, sometimes in a pejorative manner — it was something more akin to myth and art. Yet, inasmuch as this mythic picturing considered life in its wholeness, it represents a seed-image of a true organic science.

Theoretical science was to conclude that the Elements are indemonstrable and therefore of no reality. Indeed, as the eminent twentieth-century French philosopher of science Gaston Bachelard has commented, only when science completely abrogated its connection with the tradition of the Elements was the first step toward an objective (analytical) science made.[50] Science has come to call an "element" something quite different from the earlier mytho-poetic conceptions; it now means the demonstrable constituents of matter which make up the periodic table. A standard modern scientific definition of an element is "any of the 105 known substances that consist of atoms with the same number of protons in their nuclei."[51] These material elements are understood to combine to form molecules and complex compounds in a way similar to how the letters of the alphabet can be arranged and rearranged to form different words and texts. It would now be considered nonsense to declare that the elements constitute the mind or soul.

It is remarkable, in view of the above statement, that it was Bachelard who devoted the larger part of his intellectual life to an elucidation of Elemental imagery through a study of the great literature of Western culture, works ranging from Shakespeare to D.H. Lawrence, to Lautremont, to Goethe. In Europe, Bachelard's writings on science are regarded as being on a par with those of philosophers of science such as Popper, Kuhn and Althusser, the latter taking over a number of Bachelard's central insights concerning the structure and evolution of scientific consciousness. It was surprising and even a cause of dismay to some when he turned to a dedicated study of the Elements — yet this becomes perfectly understandable when we consider the whole course of his thinking. It was *through* his reflection on the origin and structure of the objective scientific mind that Bachelard came to understand the power of the Elements as fundamental "dispositions of the soul." Influenced by Husserl, he adumbrated a "phenomenology of the material imagination" in terms of four Elemental modes.[52]

Bachelard saw the Elements as fundamental "organizing ideas" that structure human consciousness, noting their "strange stability" throughout the ages in both artistic and scientific endeavor.[53] In accordance with Husserl's views on intentional consciousness, he advanced far beyond the idea of poetic imagination as "mere subjectivity" (Sartre, a contemporaneous French philosopher, moved in that direction). Likewise, he overcame the view that the Elements represent "pure myth" or the imagining of a primitive or naïve mentality. Bachelard had a profound understanding of the mutuality of "what is known" (*noema*) and "how it is known" (*noesis*) in human cognitive life. He writes:

> I think I am justified in characterizing the four elements as the hormones of the imagination. They activate groups of images. They help in assimilating inwardly the reality that is dispersed among forms.[54]

Bachelard's view was that objective matter itself can be experienced as an "image-producing force."[55] This is what Bachelard was striving to convey when he said that the function of the imagination is not merely to correspond to or reflect but to "*sing* reality."[56] The Elements, as "hormones of the imagination," are the modes of intentional consciousness in which the forms of nature come to presence in quite distinct yet related ways. An Element does not constitute matter in a physical sense — it constitutes the *imaginal* dimension of matter, that which can be perceived through the active imagination.[57]

On the relationship of art to science Bachelard remained equivocal; he spoke of loving these with "two different loves."[58] We have noted that it was through his research into how science sought objectivity that he arrived at his appreciation of the ineluctable working of the Elements in the human imagination. In his early work Bachelard had traced the rupturing process by which objective scientific consciousness evolved over the last two millennia, the emancipation of consciousness from the Elemental principles previously held to be the foundation of all knowledge. But it is just this rupturing process which now enables us to grasp the Elements in a completely new way, as "hormones" of the imagination or "observational modes." Bachelard came to see how human consciousness intends the world through discrete "image producing forces." He had rediscovered, in terms of modern phenomenology, the ancient notion of the Elemental likeness of mind and matter.

Bachelard described the Elements as "image producing forces" at work in the forms of nature — but he could not make the bridge to an actual science of the organic world. A hundred years before, the philosopher Hegel had moved in that direction.[59] It was Rudolf Steiner who, in the early twentieth century, showed that an authentic life science must develop through the cultivation of artistic forms of cognition which transcend both the theoretically discursive and the poetical as such. Steiner explains how the Elements become "observational modes" or stages in a Goethean research process.[60] He declares:

> The understanding of natural science up to now has limited itself to this one element, earth, and now we must find the way back. We must find our way back through Imagination to the element of water, through Inspiration to the element of air, through Intuition to the element of fire . . . the ascent from ordinary cognition through Imagination, Inspiration, and Intuition — is fundamentally also an ascent to the elements.[61]

Earth he calls "ordinary" or "physical thinking"; this is the purely theoretical approach, the science of mechanism, developed through a logic of cause and effect. The next stage is what Steiner calls Imagination; this is a mobile, pictorial faculty, which is able to enter into the being of things and "see" their living, metamorphosing activity in a way which physical cognition is unable to do. Through Imagination we learn to "think growth." The stage after this he calls Inspiration, the capacity to "see" the gestures or dynamic qualities of living forms. This occurs through a paradoxical emptying of consciousness, the bringing about of a highly receptive condition of mind. Intuition is the direct experience of the creative and free aspect of nature, the *entelechy* or "principle of individuality," that which Goethe called the "pure" or "archetypal" phenomenon.

Mystics seek to merge their soul with the fiery creative source of all and everything, with the universal essence. They yearn for Fire alone because, from the one-sided perspective of Fire, Earth — physical existence — can appear as a dark realm of illusion and ignorance, of the ephemeral and unreal. This was Plato's view, but it was challenged by Aristotle who, with his instinctively scientific disposition, wished to understand the world through its sensory appearance. However, from a one-sided Earth perspective (which we now call the materialist outlook), Fire can equally appear as a realm of illusion, mystical feeling and fantasy, as something which cannot be verified. An authentic science of the living world must embrace the wholeness of life, the well-roundedness of the Elemental domain. It begins in the realm of objective physical phenomena (embodying the principle of Earth), but in and through this objectivity it seeks the "pure" or "archetypal phenomenon." Earth hides the spark of creation, the life essence and activity, which is Fire. The Goethean science of the wholeness of nature, which was articulated by Steiner, develops by way of an apprenticeship with the world of material things (through Earth or physical thinking). It then traverses the Elemental realms of Water and Air through the developed faculties of Imagination and Inspiration and moves toward the archetypal realm, or Fire, through the organ of cognition he calls Intuition.[62]

CHAPTER 2

THE ELEMENTAL MODES OF COGNITION

Earth Cognition — Physical Thinking — the Mechanical

As human beings, we find ourselves in a world of discrete, particulate entities, which we refer to as "the external world." There is "I" and there is the "world of things" — and all these things appear to confront us, even to oppose us. But this is not just the experience of things as separate from and alien to *oneself* but of things as separate from each other. Earthly entities altogether have the character of being *external,* existing in three-dimensional space. R.G. Collingwood, following Hegel, describes nature as "a world in which everything is external to everything else."[63] This world of things is infinitely complicated, indefinitely multiple. The experience of the world in its externality is, in the first place, that which stimulates the sense of wonder and the will to understand. Thus we can speak of the first stage on the path of knowing: it is the primordial human experience of the world of objects, which was traditionally called Earth.

Entities in our environment have a physical presence and self-contained character. We call them "real" because they have a *definiteness*; we bump up against things literally, or they meet our senses in manifold other ways. Even evanescent and less-than-solid phenomena such as running water, the metabolism of organisms, clouds and the air, have a certain definiteness or physicality, the quality of being established in themselves — an "otherness." Most expressive of this quality, however, are the rocky elements of a landscape, that which we call the physical earth or ground, the realm of inorganic or dead substance, which is hard and unyielding, opaque and apparently motionless. It was through the everyday experience of the physical earth that the ancient philosophers arrived at the essential qualities of the Element Earth as being dry, cold, dark, separate, particular, solid or firm, highly condensed.[64]

Bachelard speaks of Earth as an "imagination" of force, of volition. He writes:

> [Earth] is characterized first by resistance The resistance offered by terrestrial matter . . . is immediate and consistent. It soon becomes the objective and honest partner of the will.[65]

Opposition, otherness, excites human imagination and energy. As Bachelard observes, it stimulates the will to penetrate and understand; it arouses courage and prevents stagnation. The quest for knowledge of the external world, for illumination amid the opaqueness of earthly substance, can thus be understood as the force of volition which develops from the experience of opposition, from the sheer otherness of things. That opposition is the primordial human experience of freedom; separateness is the condition for objectivity but also for the kind of love which can grow *through* objectivity. On the one hand there is the will to overcome which arises from resistance, out of a consciousness of freedom and separation, but this goes together with the desire for intimacy, to enter into the interior of things and find the "seed" of their internal form. This desire is the "attempt . . . at touching the grain of substances" as Bachelard puts it.[66] What we conventionally call "science" has its roots in Earth experience and it pervades all its activities; for, as Bachelard shows, our life of ideas is born of imaginal depths, out of the pulsations of Elemental archetypes.

Gathering together these indications, we can now ask: What is the thinking that is commensurate with the imagination of Earth? "We see Earth by means of Earth," declared Empedocles more than two millennia ago. The imagination of Earth pervades the thinking that has its origin in the volition aroused by the primordial experience of nature in its "otherness." This thinking meets and experiences this otherness through itself becoming objective, secure and strong — it is in this sense that we "see Earth by means of Earth." This thinking separates the perceived world into discrete objects that can be counted, measured and weighed. The scientific ideal of certainty or provability, of objective knowledge, of the exact quantification of nature — the requirement to establish facts and theories which can act as reliable foundations for human progress — expresses most purely the aim of a thinking that has become as Earth in order to know Earth.

The aims, methods and structures of objective scientific thinking, as well as the manifold "anti" approaches which seek to make the human subject the main object of thought (which point to the "constructedness" of knowledge), are all pervaded and structured by the imagination of Earth.[67] It pervades traditional metaphysical thought too, in that all higher realities have tended to be expressed in terms of Earth experience; thus we find the imagery of God as *ground* and *sustainer*. Terrestrial matter is experienced as the basis for things — it is the solid ground beneath our feet, the container of the Earth's waters, the matrix that supports the existence of all living things. Earth is the experience of ground as such. The thinking which aspires toward the well-defined, fixed, unassailable or irrefutable, is in a sense a *mimesis* of the solid rocky foundations of the Earth. Henri Bortoft refers to this thinking as the "solid world mode of conception" and writes:

> [The] very image of a separately existing world, indepen-
> dent as such of our knowing it (and yet appearing just as
> it is when we do know it), is itself an instance of the "solid
> world" mode of conception. What is seen cannot be sepa-
> rated from the way it is seen: The solid world is the cog-
> nitive correlate of the solid world mode of conception.[68]

It follows that we cannot stand outside this correlation by asserting ever more vehemently the separation of subject and object, *for this assertion, too, is precisely itself an instance of the "solid" mode of thinking.*[69]

In the modern philosophy of science, two scientific strategies are sometimes thought to be at odds with one another — *realism* and *empiricism*. It can be countered that *both* realism and empiricism are of the nature of Earth thinking. Empiricism claims that what is regarded as true and real depends upon the certainty and reliability of sensory experience, that what is verifiable is that which is experienced through the senses rather than through the mind speculatively. The realist view is that the real does not *depend* upon the human senses or consciousness at all, even if the gathering of sense data and analytical thinking are part of the scientific experiment. Scientific realism holds that things exist in reality over and beyond what humans perceive them to be, that reality is self-subsistent and "strongly objective."[70] In the annals of science there is a famous encounter which is most telling with regard to the "solid" nature of scientific realism: Dr Johnson, on hearing the view of Bishop Berkeley that reality is dependent upon mind, is said to have struck his foot with great force against a large stone and declared, "I refute it *thus*."[71]

Much philosophic effort has been devoted to weighing up the pros and cons of the realist and empiricist positions without bringing light to the fact that *both* demonstrate the same *gesture* of Earth thinking, its solid character. The methods of empiricism and realism seek to provide a sure foundation for objectivity and the *discovery* of truth (and so a solid foundation for human culture). What links realism and empiricism is the imagination of the world in its *otherness* or *externality*. In both cases the aim is to show that reality, or truth, can only be determined by excluding everything that has to do with human subjectivity, by making our thinking adequate to the character of the object, i.e., firmly established in itself. The realist notion of "truth as correspondence to the real" expresses the ideal of otherness and externality in its essence. The correspondence theory of truth and its associated epistemological problem are both bound up with an Earth form of thinking; they are expressions of the "solid world" mode of cognition.[72]

As forms of Earth cognition, both realism and empiricism — as interpreted within conventional scientific practice — rest upon a quantitative methodology.[73] Whether to furnish the hard data upon which inductive reasoning can operate (empiricism), or to assess the validity of theoretical conjectures (realism), what is required is that the external world be in some way *measured*. Measurement is the method whereby thinking makes itself adequate to the world of particulate entities; that is to say, the act of measurement is to make something definite by separating it into numerical units. As Heidegger has shown, the Greek word *theorein* was translated by the Romans as *contemplari*, which means to partition something into a separate sector and enclose it; it relates to the Greek *temnein*, to cut, and the uncuttable is *a-tomon*, atom.[74] The theorizing intellect distinguishes the separated "atoms" of experience through measurement, and the "atom" of measurement is the mathematical unit. A unit is a self-subsistent whole, which is distinct from and external to the next self-subsistent whole. The unit of measurement is thus the foundation of all Earth methodologies, the essence of all solidity. It is the very epitome of the cold, hard fact. A unit of measurement has no inner volition, no will or purpose, no soul and no capacity for self-movement; it is used to determine what is objective in the world. The unit is the imagination of solidity and otherness rendered as an abstract concept.

It was in geometry, passed down from Egyptian to Greek culture, that measurement first became a precise procedure, and geometrical construction then became the basis for the evolution of Western scientific thought. As Henri Bergson has observed:

> [T]he human intellect feels at home among inanimate objects, more especially among solids, where our action finds its fulcrum and our industry its tools . . . our concepts have been formed on the modes of solids [and] our intellect triumphs in geometry.[75]

Euclidian geometry developed from the idea of the unit or atom and is essentially a geometry of solid bodies. Joining two points, we form the line from which we then can derive the circle, sphere, and so on, in such a way that all three-dimensional forms are reconstructed and made intelligible. Aristotelian mathematics developed in relation to this geometric thinking and is characterized by its solidity. When something is measured and abstracted into numbers, the numerical result is separated from the incidental properties or qualities of the object, which, for the purposes of clear intelligibility, are filtered out.[76] Through measurement we render quantitative what is of a solid and material nature; and what is alive and growing is turned into what is hard, fixed and definite for the purposes of comprehension.

What is external has the character of being *extensive* — what is inner and of a soul (intensive) nature cannot be measured. To be extensive is to be of the nature of surface — to be solid means to occupy space and to become present to us as a surface phenomenon. In terms of practical science, the insides of things — rocks, plants, animals — must be broken open, dissected, scanned with an X-ray machine or some such device in order to be observed and measured; thus, what is inside must always be penetrated and established as the external. The method of quantitative science is to break things open — but when we do so we do not see inside at all, but only more surfaces, more visible parts; and when these are opened the experience is repeated. The principle whereby something must be externalized to be thought is revealed in the word "explanation," which derives from the Latin *ex-planare*, suggesting an outward flattening, rendering as surface — onto the plane of the external. The mathematical unit or expression of logic has no innerness or mysterious depth; it is entirely explicated, which means for Earth thinking that it is *rational*. Thinking may confront the most complex entities in the world, but to the extent that they are reducible to units they are explainable.[77]

Cause and effect relationships are the measurable relationships between particulate entities with which the science of mechanics deals. Causality is the name Earth thinking gives to the process which works between physical entities, for there must be a discrete entity or agent which is the cause and another (external or "other to" the former) which is the effect. An example is the throwing of a stone: a discrete or measurable force of both propulsion and gravity (the hand and the Earth as causal entities) leads to a measurable (and thence predictable) effect — the stone's trajectory. In this way the form of the stone's movement is said to be theoretically proven. The entities involved and their relations are *instances* of a universal mechanical law, which is beyond or separate from any particular entity or instance.[78] For Earth thinking, the ideal entity is the atom, and the ideal process which works between separate atoms is a cause and effect relationship.

Conceptual thought realizes its affinity with the world of mechanical relationships through its logic, which is the basis for any scientific argument or proof. When Bergson writes that "our logic is, pre-eminently, the logic of solids," he means that we have developed our logical processes and structures as a mimesis of the causal relationships between inanimate or unorganized entities.[79] The terms of a conventional logical structure do not evince the organic unity of an entity that is growing and differentiating; they express the causal relationships of a mechanism in which the parts merely act on each other in lawful and predictable ways. The mechanical nature of logic is apparent in the character of the syllogism which is the ideal expression of deductive reasoning: if the premises *a* and *b* obtain, then the conclusion *c* follows necessarily. The syllogism has the structure of a cause and effect relationship: two separate and irreducible terms, which are entirely external to one another, act on each other to produce a result (although this action is, of course, not physical but logical). The syllogism is a "sort of conceptual mechanism," as one writer puts it.[80] Upon the foundation of deductive logic science sets out to explain all causal processes in nature.

Turning now to a specific example of how the solid mode of scientific conception can be applied to a living organism, let us consider a dicotyledonous plant: *Sisymbrium officinale* or hedge mustard, a common European and North American species. This mode begins with careful observation of the plant as an objective phenomenon — that is, with the plant experienced in its "otherness." This is the accurate description of its physical form, and for this purpose drawings of the plant are commonly made. Such scientific illustration is a highly realistic form of visual art.

Along with a scientific illustration of this kind generally goes a written scientific description, which is a scrupulous account of the plant according to observations made with the naked eye or perhaps with magnifying instruments. Thereby a clear, objective picture of the phenomenon can be arrived at, this picture forming the basis of all further scientific enquiry concerning the morphological character of the plant.

The species *Sisymbrium officinale* is an erect annual or biennial, 5-90 cm in height, growing from May to September throughout Europe (also widespread in North America as an introduced species). The leaves at the bottom of the plant are deeply pinnately lobed with a terminal lobe that is larger and rounded. At the top of the plant the leaves have an arrow-shaped terminal lobe with one to three oblong-shaped lateral lobes. The racemose flowers are very small (about 3 mm across), and pale yellow in color. They first appear in dense clusters on short stalks, which soon lengthen into long, narrow spikes. The flowers are bisexual; each flower has four sepals and four clawed petals diagonally disposed (i.e., forming a cross — hence the name of the family Cruciferae). The flowers normally have six stamens, four longer than the outer two, and a superior ovary. The scent of the flowers is not strong, but volatile and sweet. The distinctive fruit is a bilocular capsule, 1-2 cm in length, which is held against the stem.[81] The seeds germinate quickly in spring and early summer and often achieve several generations within a year.

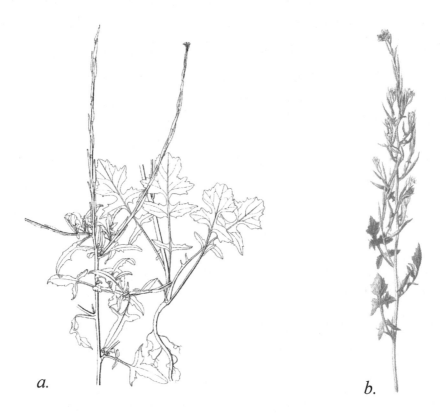

Fig. 1. Hedge mustard — scientific illustrations
a. basal rosette with root (right) and stem segment (left); b. inflorescence

Hedge mustard belongs to the Cruciferae family, order Rhoeadales. The family is also known by some as Brassicaceae.[82] This family is made up almost entirely of annual, biennial or perennial herbs and is a large family of 350 genera and about 2,300 species, distributed mainly in the cooler regions of the northern hemisphere. The genus *Sisymbrium* is made up of 80 species, including *S. officinale*. The family is characterized by features that include the following: the leaves are normally alternate and simple, often lobed and sometimes lyrate;[83] the small flowers have four petals and are usually white, yellow or orange (in a few cases red, blue or violet). The family has great economic importance as it includes food crops such as cauliflower, broccoli and turnip (all from *Brassica*), watercress (*Nasturtium*), and oils — in particular mustard and oilseed rape (*Brassica rapa*).

The wild members of this family thrive in habitats where most other plants cannot — in poor, dry grasslands, in steppes and deserts, by the seashore, in the far north, in rocky mountain regions, among rubble and in wastelands. This is why numerous species in the family are termed "weeds." The prolific foliage quickly moves to the flowering stage, with new flowers continually and abundantly produced and quickly passing to the fruit stage so that flower clusters are mixed with pods. Abundant flowering leads to prodigious seed production. In short, this plant family is characterized by its great vigor and vitality.[84]

The essential quality of the Cruciferae can also be experienced through the sense of taste. This quality is most distinctive in the taste of the cabbage and its vegetable cousins, and it appears concentrated in mustard. Mustard is obtained from the ground seed of *Brassica juncea* and other species. Sulphur-containing glucosinolates are the precursors of the mustard oils glucosides, and are responsible for the pungent taste of most crucifers.[85] Sulphur tends to accelerate metabolic processes, and its presence in the Cruciferae is consistent with their vitality and morphological dynamism.

What we have above is a standard botanical description supplemented by results of taxonomic and chemical analysis. A sharp observer, however, will soon note the limitations of such a description. Even plants of the same species can greatly differ, and their differences increase under varying environmental conditions. Understanding a species or family as a whole therefore requires a certain fluidity of thinking — an ability to perceive one typical gestalt or type in myriad forms — a capacity that goes well beyond the rigid definition of Earth thinking. We will consider this capacity more closely when we discuss the Cruciferae in the section on Air cognition (p. 55-56). Though every taxonomist uses a more holistic mode of thinking to group plants or animals, it remains an unobserved element in the cognitive process. The results are then framed in terms of Earth thinking (as rigid definitions).

After Earth thinking has separated, defined, and detailed the various aspects of a given natural phenomenon, it proceeds to explain them in terms of cause and effect. Goethe indicates that the careful observation and analysis of any natural phenomenon can lead us quickly to concepts or theories concerning its parts and relationships. The problem, he says, is that these theories, even if very clever in formulation, tend to fulfill the expectations of the scientific community rather than to do full justice to the phenomenon in question. The history of science, he says, teaches us the difficulty of seeing a phenomenon in terms of itself rather than from the standpoint of convention.[86]

We may take as an example the relationship of plants to light. People have always known through simple observation that plants grow only in conditions of light. At some point in the evolution of biological science the hypothesis was brought forward that the leaves are the points of light reception and that light *causes* the plant to grow; this hypothesis was tested and verified and has become a common classroom demonstration. The important thing to note here is that Earth thinking always frames its hypotheses in a particular way. According to the fundamental disposition of the "solid" approach, it takes the light to be one thing, the leaf another — that is, they are seen as external to and separate from one another. It then seeks to deduce the causal relationship between them. However, this is by no means the only possible approach — it is a disposition of thought evolved from long human experience in the world of interacting solids. This disposition assumes that there is a logical (causal) relationship between the light and the leaf which an experiment can reveal in exact terms. Through such experiments biology has developed and continues to refine its knowledge of the complex mechanisms of photosynthesis, cellular respiration, amino acid production and so on. The aim is to eventually elucidate all the mechanisms by which a plant metabolizes and grows.

The assumption of this Earth approach is that *everything* about the form and function of plants can be accounted for in the same way. It proceeds on the basis that, given time and patient experimentation, all the complexities of plant mechanism will be unraveled. Neo-Darwinism, for example, sets out to explain the origin of form in plants in terms of a view of evolution which also sees everything in terms of cause and effect relations. Although Darwin's hypothesis is continually being redefined, with many voices dissenting from the "classical" model, the essential thing to note here is the way all such positions frame the entire phenomenon of earthly evolution within a mechanistic conception. Mechanisms of environmental selection cause one new plant entity to take hold, another to die out. Environments act on plant entities through cause and effect processes, which are, in principle, entirely explicable. Mutants arise through modifications of the genetic structure, which are, again, explained as cause and effect processes occurring among discrete molecular entities, processes which are ultimately mechanical in nature, which have a basis in logic and can be unraveled, given sufficient time.

In truth, Earth thinking *cannot* think growth, *cannot* think evolution. Earth thinking is not commensurate with the phenomenon of growth, is not *like* growth. Growth and evolution are continuous and whole; they are not "particular" in the sense of being composed of parts into which they can be broken in order to investigate how those parts act on one another causally. Growth means that a thing grows or evolves beyond its original form; but the earlier does not *cause* the later to be. Earth thinking is not deficient in its explanatory power; rather, it is commensurate only with what is mechanical in nature where its explanatory potential is unlimited. But growth and evolution cannot be understood through the process of logical thought, which divides an organism into a complex of parts that are causally linked. A science which can "think growth" requires a dedicated disposition of thought and a particular experimental approach. An authentic science of living form will require a different kind of thinking.

Earth thinking works legitimately in the observation, description and elucidation of mechanisms in nature but overreaches itself when it seeks to squeeze the whole of nature into a single frame of reference. A living plant, animal or human being is more than an object which is being pushed, pulled and transformed by physical forces. An organism is not *essentially* a machine — even an extraordinarily complex one — and it is this "moreness" which the Goethean phenomenologist wishes to bring to thought. Specifically, what is being sought is the manner in which a living essence or "body of formative forces" (*entelechy*) creates and governs the purely physical nature of a living thing.

We would never go beyond an Earth knowledge of the world were it not for the power of reflection to perceive the nature and limitation of this mode of conception.[87] An immediate way this limitation shows itself is in applied science; here it is apparent that many of the technologies created by this kind of thinking wage war upon the living being of nature and have brought about an ecological crisis. Only through reflection is it possible to see how science has come to promulgate a world-view that denies the "moreness" of organisms through the presumption that analytical intelligence represents the only way of knowing.[88] Henri Bergson puts the case succinctly:

> [O]ur thought, in its purely logical form, is incapable of presenting the true nature of life, the full meaning of the evolutionary movement. Created by life, in definite circumstances, to act on definite things, how can it embrace life, of which it is only an emanation or an aspect?[89]

To move beyond logical, analytical thinking can appear as an enormous risk to the scientific endeavor, a leap into the unknown. But for the scientist, on a personal level, it is a moral deed, an act of relinquishing an habitual frame of reference in order to approach the essential nature of life. Elementally, it is the metamorphosis of Earth to Water.

Water Cognition — Imagination — the Sculptural

Earth thinking and its associated methodologies have given rise, over many centuries, to a family of sciences which includes botany, zoology, biochemistry, ecology, genetics, microbiology — collectively named the "life sciences." Following from the arguments of the previous section it may now be asked whether these represent genuine life sciences at all. Indeed it is the case that, across the whole span of the so-called life-sciences, the phenomena of nature are being reduced to the terms of Earth cognition. No matter how ingeniously science seeks to penetrate the complex interactions and processes of the organic realm, all such theorizing is in the manner of an Earth logic. A whole history could be written concerning the creation of new sciences by Earth thinking, the aim of such a history not being to negate the value of these sciences and their complex evolution but only to see them more clearly for what they in reality are — both their virtues and their limitations. The true "turning" comes with the recognition that life is more than Earth thinking will *ever* have the potential to comprehend, that there is a Water thinking which is of an entirely different nature.

Bergson has described the action of mind whereby thinking comes to an authentic experience of *duration*, the continuity of movement and transformation which characterizes living things. He attempts to describe the movement of this thinking, which, through its own act of will, is learning to go beyond itself, and he does so through the image of the transformation of walking to swimming. By walking he is referring to the form of intelligence which understands itself through "the resistance of the solid earth"; by swimming he means entering a new medium that requires the mind to "get used to the fluid's fluidity."[90] Bergson is accurate in describing the activity of this thinking as being of a fluid nature — here it will be called "Water cognition."

What is alive has the quality of being "for itself;" that is to say, it is not merely subject to external agents acting upon it. It was clearly enunciated by Kant, Schelling and Hegel that organisms cannot be wholly explained according to the mechanistic logic of cause and effect, because organisms come forth out of themselves. The movement which we call "growth" is not the same as the movement of one atom colliding with another or a hand throwing a stone; a plant emerges out of and for itself, not merely because of forces acting upon it. The questions for the life scientist become: how can we authentically think the becoming of a plant? What is the lawfulness that unites plant parts in an organic unity?

The pre-Socratic thinkers recognized in the Water principle the character of being transparent (relative to Earth), fluid, penetrating, surrounding, formless and continuous, whole (as opposed to what is separate or particulate).[91] We can go further with the help of science and describe water as the substance indispensable to life. It has three prominent characteristics: it is a great solvent and readily receives other substances into itself through the process of dissolution. It tends toward rhythmic motion and is associated with all rhythmic process in the life realm. And it is a highly sensitive medium, having no definite form of its own and, for this reason, being receptive to the form and motion of other entities — embracing them, coating them, responding to their movements.[92]

Gaston Bachelard has evoked many dimensions of the imagination of Water, one of these being the experience of working with soft pastes (*pâtes*), which are formed when water mixes with earth. A sculptor working with clay experiences *through his or her hands* the inner continuity of form; through modeling we experience duration as an elemental force rather than substance that is perceived visually. Bachelard writes:

> This duration . . . is not *formed*. It does not have the different resting places provided by successive stages that contemplation finds in working with solids. This duration is a substantial becoming, a becoming from within.[93]

This sculptural modeling work gives rise to an intimate imagination of fluid substance; it is

> a medium of energy and no longer merely of form. The dynamic hand symbolizes the imagination of force.[94]

Bachelard describes the working of soft substance as an imagination that

> is rhythmic, with a heavy rhythm, that takes hold of the whole body. It is thus vital. It has the dominant characteristic of duration — rhythm.[95]

The modeler is *Homo faber*, the human maker whose activity arises from an imagination of Water. The actualized form of things is grasped through Earth perception; however, as Bachelard shows, through the imagination of Water and the experience of working with soft substance, what is grasped is the rhythmic arising of form, the *activity* of formation rather than the formed *results* of that activity. In short, Earth thinking is spatial in nature (even in its conception of time); Water thinking is temporal (time as experienced continuity, duration). [96]

These evocations of Water help to awaken the organ of cognition which will here be termed Water thinking or Imagination. Goethe recognized that there is a power of mind which is like the fluid becoming of nature and he called it "exact sensory imagination," advising:

> If we wish to arrive at some living perception of nature, we ourselves must remain as quick and flexible as nature and follow the example she gives.[97]

Water thinking, Imagination, is the way of knowing which belongs to the "what is known" which is growth. We have seen that cause and effect (Earth) thinking seeks the law (explanation) of the form of leaf and flower in an evolutionary mechanism. Water thinking is the thinking which understands growth (evolution or "substantial becoming") on its own terms because it is *like* that becoming. In the plant "substantial becoming" or growth is what creates the different "resting places" or "successive stages" — namely, the leaves that appear rhythmically up the plant axis.

The Goethean biologist Jochen Bockemühl has described how this "exact sensory imagination" is practiced: he writes that, just as water "molds itself to its surroundings or tries to form an even, all-embracing surface," so an exact imagination "molds itself to one form and then dissolves it again as it flows on toward the next form."[98] Like water, imaginative thought is sufficiently plastic and sensitive to take on the form of another being. Water thinking "runs through" the forms of the leaf, the flower, then flows into the forms of the fruit and seed. With our exact imagination we enter into the leaf shapes and move between them, through their sequence of growth. We feel our way with the sensitivity and formlessness of water into the precise configurations and qualities of the physical form. In Water thinking, we learn to "dwell" imaginatively in the form of living beings with a thinking that participates rather than remains as the external observer. This participatory Water mode of thinking, Bockemühl indicates, "leads us not only to the surfaces of things, it unites us with their processes."[99] Michael Polanyi has called this participatory activity of thinking "tacit knowing." As he puts it, we no longer remain external to objects but rather "we pour ourselves into them and assimilate them as part of ourselves."[100]

Let us take the plant we were previously considering, hedge mustard — but this time just observe the changing form of the leaves up the axis of the plant.

The exact imagination flows into and between the forms along the series, forwards and backwards, molding itself ever more exactly to the forms and seeking the law of their continuity. Gradually the imagination develops an understanding of how one leaf form evolves into another in the growth of the plant, the law of their "belonging together."

In this way we learn to think the growth of the plant. The changing form of leaves is a good place to start with this work because the transformations are relatively small and a whole series can present itself on a single plant. But a developed Water thinking needs to account for the whole form of the plant. The movement from leaf to flower represents a more radical aspect of the becoming of a plant, but the exact imagination works into and through this progression in the same way. Through leaf progression to flower it may thence move to fruit and seed, backwards and forwards through the whole cycle of growth. What is discontinuous to

Fig. 2. Hedge mustard — showing changing leaf forms

immediate perception (leaf-flower) is gradually revealed as a continuity in the fluid movement of organic becoming. Water thinking embraces and unites all the structures which have been identified in the Earth stage of the work.[101]

Through a dedicated and lengthy practice of "exact sensory imagination" we may gradually develop a comprehensive inner picturing of the total coherence or wholeness of a plant. This coherence is seen by the "eye of the imagination" and judged through what Goethe called the power of "beholding thinking" (*anschauende Urteilskraft*).[102] This is not a logical judgment for, as we have seen, we are not at this stage dealing with cause and effect relations. The leaf does not *cause* the flower, yet leaf and flower may become entirely intelligible in terms of each other through imaginative participation in the growth process. The judgment made here is mathematical in *style* but not in *content*; that is to say, the *exactitude of the method* of relating forms through "beholding thinking" can be called mathematical even if it is not dealing with number (i.e., quantities).[103]

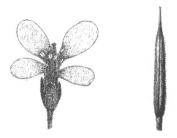

Fig. 3. Hedge mustard — flower and fruit

The lawfulness that unites leaf and leaf, leaf and flower, is not *causal* but *formative*. Ronald Brady has described the separate leaf forms, as in the above example, as the "gestalts" or static "moments" that are generated out of the continuous "moving" form. He writes:

> It might seem counter-intuitive to speak of movement, rather than an object making the movement, as generative, but between the forms and their movement there is only one possibility ... the movement is not itself a product of the forms from which it is detected, but rather the unity of those forms, from which unity any form belonging to the series can be generated.[104]

One leaf does not physically transform into another; we cannot physically *see* formative movement. What endures from one form to the next is the generative "form" itself. Brady argues that through the mobile imaginative thinking which thinks *as* the moving form, we "see" all successive stages (i.e., leaves and other organs) *as the same form*.[105] This is what Goethe meant by metamorphosis. The authentic science of living form, at this second stage of Water thinking, is cultivating the perception of the formative activity that results in the metamorphosis of form. Water thinking could also be called a "sculptural thinking" — it sees and understands how each physical form is sculpted out of an entirely plastic formative movement.

Let us now return to the Cruciferae family. The attempt to encompass this extraordinarily diverse plant family in thought calls for the same mobility of thought required to follow the metamorphosis of a single plant. As we compare and move from species to species with this mobile, sculptural thinking, we see how the creative potential inherent in this family pours itself into particular organs: into the base of the root in the rutabaga, into the leaves in the cabbage, into the main stem in the kohlrabi, into the buds in Brussels sprouts, into the inflorescence in broccoli and cauliflower, and into seed production in mustard or in such wild forms as hedge mustard and pennycress. Transforming one of these plants into another in exact imagination is a wonderful exercise in sculptural Water thinking.[106]

Sensible form (shape or structure) should not be confused with form as non-sensible formative movement. This second notion of form as movement or "substantial becoming" is in accordance with Aristotle's understanding of form as *entelechy*. As Jonathan Lear writes of Aristotle's teaching on this subject:

> One cannot . . . identify natural forms with an organism's structure. Structure helps to constitute the form, but forms are also dynamic, powerful, active. They are a force for the realization of structure.[107]

Aristotle recognized that organisms have within them their own principle of self-movement and self-generation. This is universal life become concrete or incarnate as individualized life or organism. The life-principle of an organism, the *energeia* or *entelechy*, he considered to be the actual nature of the organism, its living idea. *The entelechy is not an explanatory idea;* it cannot be logically thought out and argued for. Rather, it is a living, active idea, which can be cognitively *experienced* by means of the active imagination.

Purely intellectual thinking, Earth or physical cognition, is mechanical and dead, and this is precisely what allows it to apprehend what is dead and mechanical in nature. Rational consciousness gains its clear reflective capacity and objectivity through a death process, through the separation of thought from the vitality of subjective feelings and impulses of will. This separation gives rise to the experience of freedom, without which there could be no search for knowledge. Earth thinking could also be called "head thinking" — rigidly logical, abstract and divorced from the rest of the human organization. Water thinking means that an aspect of the middle part of the human being has become active in the cognitive process. This is the mobile quality of feelings, the capacity of human sentience to move out and enter into other beings in nature, to sensitively live into and participate in their growth process.

The conventional scientific view is that feeling impinges upon or sullies thinking — and this is certainly true in relation to purely logical and mechanical thinking. But we have seen that such logical thinking is not adequate to the growth process in organic form. With Water thinking or Imagination we are

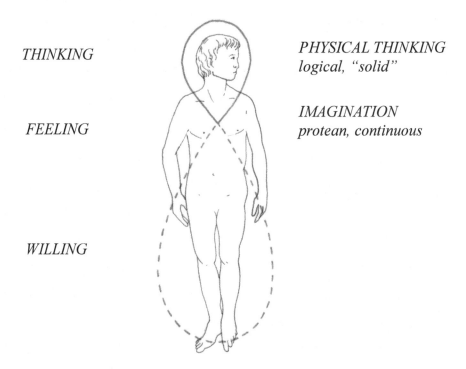

THINKING

FEELING

WILLING

PHYSICAL THINKING
logical, "solid"

IMAGINATION
protean, continuous

Fig. 4. The three-fold human being — Imagination

literally *thinking with our feeling* — but here we are not speaking of a personal content of feeling or emotion. What is meant is that a certain aspect or *capacity* of the feeling life is intensified and heightened into an organ of cognition. In relation to Water thinking this is the protean power of feeling, its character of continuity and transformation. It is the feeling with which a sculptor forms his works. This is what allows feeling to mold itself to the fluid, metamorphosing form of another being, and it is by virtue of this plasticity that we speak of Imagination as a "sculptural thinking."

Imagination, as a mode of thinking with a definite methodology, is still largely a potentiality in the sphere of the life sciences. Yet, even though it represents a very large step beyond the conventions of scientific practice, Imagination has its limit as a mode of cognition. We have seen that it is through reflection that Earth thinking grasps its own limitation — and so it is with Water thinking. Only through the power of reflection can thinking go beyond itself, can it awaken to further possibilities of itself as dictated by the living beings it is endeavoring to understand. This limit to Water thinking is the perception of *entelechy* as plasticity, form as formative movement. The organism is more than just formative movement, for within every formative movement in nature there is a definite *gesture*. Thinking needs to pass through the sphere of plastic form into the more ideal or supersensible sphere. Elementally speaking, this is the metamorphosis of Water into Air.

Air Cognition — Inspiration — the Musical

By learning to participate imaginatively in the "otherness" of the phenomenon we begin to experience formative movement; this is the metamorphosis from Earth to Water. We start with the methods of an exact empiricism, but by the Water stage we have advanced to an entirely different form of thought — a sculptural thinking which can be cultivated through modeling work. That which began as a clearly recognizable scientific process of observation and measurement has begun to turn into something more like art. Yet — and this needs to be emphasized strenuously — art has not been drawn into this process of investigation in an arbitrary way. The necessity of art is dictated by the thing itself — in the first place, by the perception of the formative, plasticizing Water element. Science becomes artistic out of its own requirement to realize itself as a true science of living form. In the next stage, the Air stage, we enter the "musical" dimension of the organism, meaning that we learn to perceive formative movement as *gesture* — as experienced, for example, in the dynamic qualities of melody and musical tone.

We have already considered that volitional force, the will to know, to penetrate and understand, has to do with the *oppositional* nature of Earth. Earth is experienced as an awakening force, arousing strength, courage and the experience of freedom, even the desire to master or dominate if this impulse is allowed to become one-sided and exaggerated. Water already represents a mediating, softening influence; by imaginatively flowing into things, we are, in a way, giving ourselves over to them for the sake of letting them appear just as they themselves are, not according to a willful desire to conquer through our knowing. With Air we move beyond opposition, for we are moving beyond materiality altogether. In relation to this Element, Bachelard writes that "[t]he substantial imagination of air is truly active only in a dynamics of dematerialization . . ."[108] Air is experienced as the void, the spaciousness or nothingness which, in the way Bachelard evokes it, is not merely an absence of matter but an *overcoming* of matter. This very "nothingness" now becomes a condition of the process of cognition, and it is through a "dynamics of dematerialization" that the organ of cognition named Inspiration awakens.

Henri Bortoft has described a decisive moment of transition in Goethean research as a "subtle reversal of will":

> [T]he researcher, in directing attention exclusively to the phenomenon, is in fact surrendering to the phenomenon, making a space for it to *appear* as itself. This provides the condition for the reversal of will to happen, from active to receptive will, whereupon it is the *organizing principle* (which is the necessity) of the phenomenon itself which can come to expression in the researcher's thinking.[109]

"Receptive will" occurs through a conscious participation in the phenomenon, already developed in the Water stage, in a way that allows researchers to "offer their thinking to nature so that nature can think in them."[110] Bortoft calls this the third state of thinking, no longer active and assertive, yet not merely passive; "receptive" meaning here the condition of mind which is the reconciliation of activity and passivity.[111] These notions of "reversal of will" and the "making a space in thinking" are keys to understanding how the Air cognitional attitude awakens at this stage of the research process.

"We see Air by means of Air," declared Empedocles. The ancients viewed Air as being related to Water in its fluidity, yet lighter and almost immaterial. They saw in its tendency to rise and expand a closeness to Elemental Fire; in the vast and luminous space of the upper atmosphere which they called the "aether" they experienced the Air principle as the clarity "which makes things visible."[112] The cognitional attitude which may be called Air is the light-filled spaciousness or transparency of mind which allows a meaning to be seen *through* a material form. Bockemühl calls Air "the element of acquiescence" and writes that

> it is characteristic of air to expand in all directions, offering its own being and activity in order that the being and activity of another can appear. Insofar as we move inwardly in accordance with this image of the air, we reach the cognitional attitude corresponding to the air element. An inner readiness is thus created for that which manifests in the world to reveal itself in us, as an image which discloses a being.[113]

This "inner readiness" is a heightened condition of receptivity of thought — we speak of Air cognition as an intentional act by virtue of which the gestural language of the phenomenon can be experienced. Air thinking can be related to Heidegger's description of thinking *(Noein)* as "authentic grasping," as a "letting something come to one," and to his discussion of truth as *aletheia* or "bringing forth into unconcealedness."[114]

Thinking becomes empty, moving beyond the posture of security and solidity provided by conventional logical structure which is inherent in the analytical method. In practice, it utterly relinquishes and releases its contents — "into the void," as it were. This appears nonsensical from the point of view of Earth cognition; it seems to represent the very antithesis of what science sets out to achieve, which is a firm, objective grasp of the world through a rigorous chain of reasoning. Yet we are not talking about negating the content of what has already been gained through the Earth and Water stages, only about the form of thinking required for these Earth "facts" and Water "Imaginations" to become transparent so that their meaning, their essential character, may disclose itself. Aristotle understood that the mind can achieve such a heightened receptive state and described this stage with specific reference to the Element Air.[115] As Aristotle saw it, in this heightened condition the mind is entirely transparent, yielding and able to receive the true form of the thing, a form not visible to the physical senses alone, yet which is not "other to" or "beyond" the physically perceived form. Indeed, the physical form is seen as its external manifestation.

In the Water stage we described the *entelechy* as becoming experienceable as duration, through the perception of formative plastic movement cultivated in the first place through the kind of modeling work which develops the faculty of fluid inner picturing (Imagination). In Air cognition, formative movement becomes perceptible as something more than plasticity. Living form is distilled in consciousness so that it now appears as a *dynamic gesturing*. A gesture is a movement through which a definite idea or meaning becomes evident. The meaning referred to here is not that which is arrived at through discursive logic. Rather, we are entering the province of artistic meaning, that which the philosopher Suzanne Langer has called "presentational" meaning.[116] Specifically, Air thinking represents the heightening of thought from a "sculptural" to a "musical" perception of form.

Water or sculptural thinking intensifies the protean character of feeling, its capacity to move into and through the form of a living being and thereby grasp the continuity of its formative movement. Air or "musical" thinking awakens when another capacity of the feeling life is intensified and becomes an organ of cognition; the expressions "inner readiness" and "acquiescence" have been used above to evoke this particular capacity. The "watery" aspect of the feeling life is always assuming one form or another but the "airy" aspect of feeling simply *makes an inner, receptive space* in which the musical gesture of form reveals itself. Thus we can continue to render diagrammatically the different Elemental stages of thinking in terms of the soul constitution of the human being.

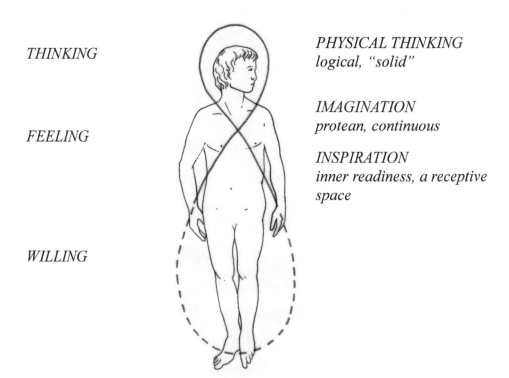

THINKING

FEELING

WILLING

PHYSICAL THINKING
logical, "solid"

IMAGINATION
protean, continuous

INSPIRATION
inner readiness, a receptive space

Fig. 5. The three-fold human being — Inspiration

Music — the most immaterial, most aeriform of the arts — is a pure language of gesture; as such, the painter Kandinsky (in his book *Concerning the Spiritual in Art*) recognized it as the teacher of all the arts. The meaning of music derives from the complex tapestry of gestures or dynamical forces which constitute its symbolic form, its language. A tone is nothing but a gesture or force — it is not any kind of solid thing which is being moved or forced. In music we hear and understand forces. The philosopher and musicologist Victor Zuckerkandl shows how meaning arises in music when tones are perceived in relationship to other tones within the dynamic field, or force field, of the octave. He writes:

> The balanced tone stands for the field's center of action; from the latter radiate the forces that act upon other tones in various ways. The dynamic quality of each tone is determined by the dominant constellation of forces at the place where it is sounded.[117]

What Zuckerkandl is endeavoring to describe, within the strictures of discursive language, is the holistic, qualitative cognition which is ordinary musical experience. A tone is not a particulate Earth entity or atom, and its force is not mechanical in nature. Zuckerkandl is saying that each tone arises from the "necessity" or lawfulness of the whole field of tones (the piece of music); it is not a sum total of measurable mechanical forces. The radiation and interaction of tonal forces is purely qualitative, but for the musical ear it is precise in its meaning.

Each tone in the diatonic scale is a condensation of the dynamic organization of the octave, a particular focusing of its field of forces. The octave is acoustically of a higher pitch than the tonic, but in terms of dynamical forces it is a return to the same — 8 is 1 for the next octave.[118] The prime, or tonic, is at rest within itself and as such is the octave's primary center of force; the step away from it — 1-2 — "runs counter to the will of the tones" to remain at rest, as Zuckerkandl describes it.[119] The intervals up to the third — 1-2-3 — speak gesturally as an "away from" the prime, remaining connected to it even as they move further away from it. What we hear is how these tones relate principally to the tonic as their center of gravity, their "past" from which they are an "away from." The fourth represents a threshold, for a new center of gravity is now established even while it still gestures something of an "away from." In the fifth we hear this new center as taken hold of by an entirely new force dynamic — the octave. The fifth gestures something of a "pointing toward."

The sixth, and especially the seventh, express predominantly a "toward" and a "yearning" to resolve in the octave — 6-7-8. The octave, not yet sounded, is exerting its influence from the future, which we hear in these tones as the gesture of "not yet." The important point is this: in the intervals we experience, and understand musically, time as it flows in two directions. In the lower intervals we hear a movement away from what comes before. But in the higher intervals the future is casting its shadow into the present. These intervals are in effect prefiguring their resolution, and this, far from being something strange or irrational, is an entirely normal part of musical comprehension.

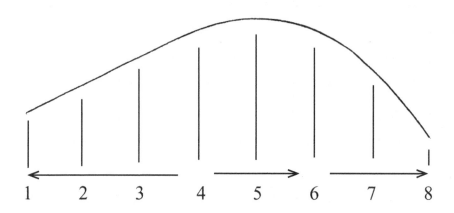

Fig. 6. Force field of the octave

The form of a piece of music comes about through an "organic" development of tonal relationships. Music is not actual organism; it is the "spiritual organic," to use Goethe's expression. Music makes perceptible the formative gestures of growth which are active in all living forms, in plants and animals as much as in the human being.[120] The different tones live and relate to each other within the octave; they focus and form the force field of the octave in different ways just as each organ of a plant — the seed, leaf and flower, for example — focuses and actualizes the formative power (*entelechy*) of the organism at different stages of its growth. Tones can in no sense be compared to the parts of a machine, which are only related through cause and effect processes. Tones develop out of each other, they grow and pass away, each focusing the total living dynamic field in specific ways at different points in time. When we speak of the musical form of a plant, what is meant, of course, is not audible tone. The music that permeates nature is a perceptible reality, but only to the organ of cognition which has been cultivated to that end. Steiner describes Inspiration as "hearing" the "toneless music."[121] In other words, this is not hearing as an auditory process; indeed, we perceive the plant principally through the physical organ of sight. What is meant is that the faculty of musical cognition — cultivated initially through the physical hearing of musical tones — becomes a "higher" capacity of the other senses as well. We see, touch, taste and smell the physical plant, and through that we come to inspiratively "hear" its musical structure.[122]

Let us return again to the plant we have been studying in the Earth and Water stages — hedge mustard. In the Water stage we "ran through" the series of its organs with exact sensory imagination, participating imaginatively in its forms, backwards and forwards, gradually deepening our experience of its growth process, specifically the metamorphosis of its leaf forms. Through Water perception, *duration* or growth becomes comprehensible; this is the sculptural perception of how each leaf is "modeled" into its form from out of a single formative movement. At the Air stage we speak of the metamorphosis of thinking from the sculptural (or plastic) to the musical. The plastic formative process becomes transparent, as it were, to its formative music, which is the actual structure of the plant's "time body" of formative forces. (This can legitimately be called a "body" since it is not a description of chance events but of a definite series of structures in time, each one being an expression of the wholeness of living form.[123])

Goethe originally spoke of the leaves of the plant in terms of expansion and contraction, away from and toward the stem or plant axis. The Goethean biologist Jochen Bockemühl has isolated four distinct leaf-forming activities within this overall expansion and contraction.[124] These activities or gestures are *stemming*, *spreading*, *articulating* (differentiating) and *pointing*. They could also be called "formative tendencies" to distinguish them from the *resultant* form (meant when we use words like "petiole" or "blade"). They do not indicate what a plant is in its *formed* nature (that is, as a physical entity) but in its *forming* nature.

Stemming is a stretching outward from center to periphery as an axial, unidirectional "push." Further up the plant axis, as the leaves become larger, *spreading* is a widening and bilateral expansion into a horizontal plane. Higher still, another activity appears to work in upon the spreading activity, as if eating away its substance, leading to a differentiated leaf-form; *differentiating* is a dividing or indenting of the leaf from the sides. Finally, the highest leaves near the flower appear to have been engulfed by this inworking activity, reducing them to small spear-shapes, which have the opposite form to the first leaves. *Pointing* is a contracting from the periphery to the center and an engulfing of the axial stretching (of the stemming activity). As indicated by Margaret Colquhoun, these four activities are not actually separate but interweave in every leaf-form, each predominating at different positions on the plant axis. All four leafing activities are clearly evident in herbaceous plants such as hedge mustard which display a whole metamorphic series. In many other plants, leaf shape emphasizes only one activity, with the other activities appearing in a more subtle or hidden form in the development of individual leaves.[125]

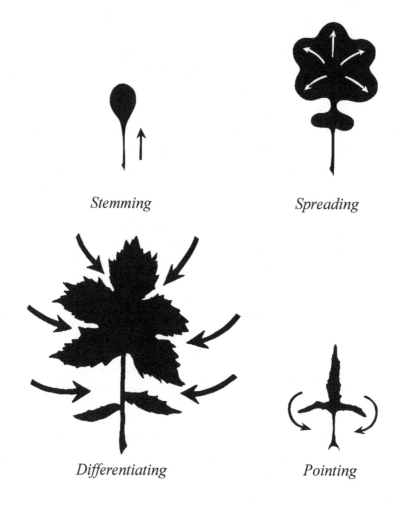

Stemming   Spreading

Differentiating   Pointing

Fig. 7. The four leafing activities of hedge mustard (not to scale). (After M. Colquhoun and A. Ewald, New Eyes for Plants, [Stroud: Hawthorn Press, 1996], p. 82.)

To speak of such particular *formative actions* means we have gone beyond organic form as plasticity (duration) and entered the realm of form as gesture or musical meaning. The terms "stemming," "spreading," "differentiating" and "pointing" are actually efforts to "say" the essential musical form of the vegetative plant; the integrated dynamic of these gestures is its musical meaning or the structure of its "time body."

In the gesture of each of these four activities, we "hear" the interweaving formative activity of the intervals of the octave; however, we should bear in mind that any particular plant species represents a highly complex and unique music.[126] These four terms indicate only the principal formative gestures of leaf growth, and to give expression to the specific nature of a plant, or to the gestural structure of any living form, we require a more developed "musical" form of languaging. This is what is attempted in the landscape study documented in Chapter 4.

The prime with its potency, yet inactivity — "the restfulness of unison with itself" — is predominately heard in the seed but also in the meristematic stem or growing point, the uppermost or outermost tips of the stem.[127] In this sense the seed-force is raised up and sounds as the prime throughout the growth of the plant as its formative potential.

In the activities of *stemming* and *spreading* are heard the "away from" gesture of the intervals from the second to the fourth, held to a greater or lesser extent by the center of gravity which is the prime. The second sounds the primary activity of growth, the commencement of movement from out of the restfulness of the prime — in the first place in the unidirectional upward growth of the stem from the seed, but then in the outward growth of the leaves, the activity we are calling *stemming*.

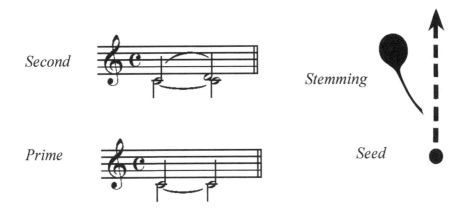

Fig. 8. Hedge mustard — prime to second (stemming)

We hear the major third in the activity of *spreading*, which soon takes hold of the *stemming* activity of the basal leaves of hedge mustard. This gives rise to the enlarged and rounded forms in subsequent leaves. The fluid, expanding quality of the major third develops nothing of the particular or the specific; the rounded leaf shape has a universal, general quality, which still sounds something of the all-potential of the seed. But, as we shall see, even in leaves low down on the plant axis of hedge mustard, *spreading* is being interpenetrated by the activity of *differentiating.*

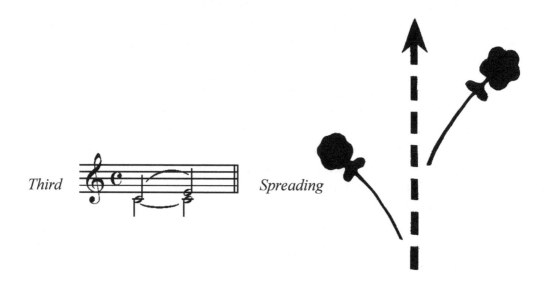

Fig. 9. Hedge mustard — major third (spreading)

The dynamic of *differentiating* already sounds in the time-body of hedge mustard in the basal leaves where lobing becomes evident. In leaves higher up, this lobing appears around the leaf margin as well as in the overall leaf-form, giving rise to the deep division of the leaf into a more or less symmetrical series of lobes. In this activity of *differentiating* we experience something of the soulful quality of the minor third; this sounds in the uniqueness of the inward-turning gesture which is active in the shaping of the leaves at this stage. In the fourth this uniqueness consolidates; something now has come into its own in the growth of the plant — a new center of gravity has been established. The leafing activity at this point no longer sounds an "away from" in relation to the prime. The fourth has the character of "waking up to itself" which, pertaining to this plant's metamorphosis, is the first coming-to-presence of the plant's species or specialized character.

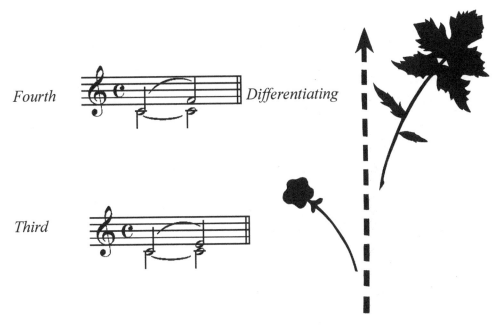

Fig. 10. Hedge mustard — third to fourth (differentiating)

The activities of *differentiating* and *pointing* live in the stream of time working from the future, the "not-yet" of the octave's field of forces. Musically, in the transition from the activity of *differentiating* to that of *pointing*, we hear the intervals fourth to seventh, balanced in the fifth. The sixth, and especially the seventh, are entirely governed by the potentiality of the future, the "not-yet-sounded octave." In the plant's time-body, the force dynamics of *differentiating* and *pointing* prefigure the flower — or, we could say, the form of the flower is "calling" the vegetative plant toward itself. What this means is that the plant's specific nature, first establishing itself in the fourth, is anticipating its own fulfillment. The leaves under the influence of *differentiating* and *pointing* are the prophetic form of the flower.

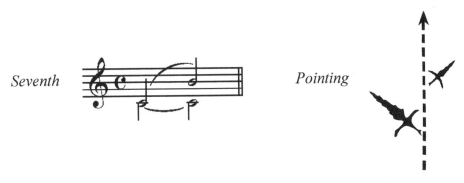

Fig. 11. Hedge mustard — seventh (pointing)

Pointing manifests itself in the middle to upper leaves of hedge mustard where the rounded lobing becomes angular, the leaf divisions gradually fuse into a tripartite leaf, and the leaf margins become serrated. This is most accentuated high up on the plant axis where the pointed leaves are held close to the stem. The fine arrow-like *pointing* gesture of the highest leaves represents an inward gathering that sounds the almost tremulous yearning of the seventh for the octave. This is the moment before the entirely new form of expansion into the flower. Here the vegetative plant is contracting to its original center of gravity — the stem — but only so that something new — the floral plant — can take form. We referred earlier to the stem with its growing point as the seed-force raised up. The stem now thickens and elongates to form the flower receptacle or torus; the leaf metamorphosis has prepared the stem for a new expansion. This activity represents a "returning-to-the-same-which-is-different," which is the gesture and meaning of the octave. The force dynamics of *differentiating* and then *pointing* have progressively engulfed the earlier expanded forms of the leaf (which are the expression of *spreading*), reducing them to a minimum, in a sense dematerializing them. The idea of the flower is asserting itself on the vegetative plant from the future, transfiguring vegetative substance, purifying it. The flower is the consummation of the idea that lies in potential in the seed and that is still more or less inchoate in the vegetative plant.[128]

Fig. 12. Hedge mustard — octave of seed to flower

The octave of growth sounds the development of the vegetative plant, from the potentiality of the idea in the stillness of seed substance to its form revelation in the luminous color, number, shape and scent of the flower.[129] This drama of formative gestures on the inspirative level —

through which substance is distilled into idea (form) — is what Goethe called "intensification" or "heightening" (*Steigerung*).[130] In relation to hedge mustard we see that the species nature of this plant is a *specific music*, a time expression, which lives in the plant in the dynamic between its organs, just as a melody lives in the "in between" of its tones. Thereby we come to understand the gestural structure and musical meaning of this plant; it is the music which sounds in what this plant goes through to realize its floral form, in how the deeply dividing gestures and rounded symmetries of its leaf-lobing condense and refine themselves before consummation in the symmetrical simplicity of its cruciferous flower.

The above is a description of what might be called the "octave of vegetation." But this octave of vegetation is only one dimension of the plant's musical structure, one movement of its time-body. If we consider the whole life-cycle of the plant, from seed to seed, we can perceive another octave of growth: the octave of the whole plant. Just as a movement of a symphony has its own structure and time body within the structure and time body of the total work, so we must think of octaves within octaves in the musical form of the plant. The octave of the whole plant accounts for the metamorphosis of all organs — from the seed, stem and leaf, to the calyx and corolla, stamens, pistil, fruit and seed. In this time structure the new seed is indeed the octave, the "returning-to-the-same-which-is-different." In this case time streams from the future, which is the new seed, drawing the *flower* toward itself, heightening and purifying it.[131]

What we have arrived at here is a preliminary Inspiration of leaf metamorphosis in hedge mustard. The four leafing gestures have been related to the intervals of the octave, which sound the fundamental gestures of the time-body of living plant form in terms of two streams of time — from the past and the future. But this does not mean that the results of all inspirative research need to be rendered in terms of the intervals or as musical notation. What is required is a language of gestures adequate to inspirative insight. As we continue to develop our understanding of hedge mustard we must intensify inspiratively everything that has been worked with at the Earth and Water stages of the research. This means perceiving with a musical sensibility not just the metamorphosis of leaf shape, but also manifestations of taste, scent, color and texture — not one in isolation from the other, but each as an expression of the wholeness of the organism which is imparting itself.

Ultimately, the essence of a plant family — that which allows us to recognize each of its specific forms as its members — is grasped as musical gesture through Air cognition. Inspiratively, each species can be heard as a tonal manifestation of the family. Thus, in the Cruciferae, what sounds through the entire family is a vitality that pours itself forcefully into different organs in different species. This vitality appears as massive substance in the cultivated vegetable species and as sheer generative vitality in the wild pioneer species such as pennycress or shepherd's purse, which grow as weeds in inhospitable areas. In the wild species the external, physical vitality sounds a characteristically intense inner dynamism. The "time organism" of these plants, dominated by stemming gestures, accelerates. In the rapid movement of growth the foliage never luxuriates in spreading activities; nor is there time for stemming forces to gather upon themselves and harden, allowing shrubs to develop. Rather, spreading is rapidly taken hold of by forces of differentiation and pointing as the plants reach energetically toward their flowers and fruits. This dynamism likewise sounds "tonelessly" in the vigor of their sulphurous taste (compare p. 33). In the *shaping* activity of these plants there is something sparse, minimal; but in the activities of transformation and substance creation the intensity is very great.

With Inspiration — musical understanding which has become a stage in the development of scientific research — we have plainly advanced far beyond the conventions of scientific empiricism and logic. Yet reflection discovers the limitation even of this highly intensified observational mode. At the stage of Inspiration, thinking is still on the way to a well-rounded understanding of living form, still in a condition of seeking, still moved to activity by the experience of the phenomenon as "other." The threshold of Air could be described as the moment of passing from the experience of a phenomenon's "otherness" to the experience of being at one with its creative essence. Elementally, this is the movement from Air to Fire.

Fire Cognition — Intuition — the Poetical

This has been the course of our pathway through the Elements: through solid Earth toward ever-active Fire. The Water, Air and Fire observational modes are reached *through* the clear observation and exact rational consciousness of Earth, and this is what makes the Elemental approach the pathway of the scientist and not the mystic. By no means have we abandoned the logic of clear scientific reasoning for a flight of the artistic imagination. Rather, our method has progressively transformed the purely sensory and mechanical aspect of the organism to reveal its creative dimension.

"We see Fire by means of Fire" declared Empedocles; only the seeing that is *like* Fire can know Fire.[132] What this means is that thinking must become free creative activity in order to grasp the living essence of form (*entelechy*), which is free creative activity. Creation is the word we have for that which comes forth out of its own activity; "self-shining" or "self-generation" is of the nature of Fire. Fire is constant activity, but nothing is *causing* that activity (as, by contrast, an action by an external agent causes the movement of earth, water or air). Activity is what Fire is in itself. Fire "lives" by virtue of the substance that it destroys, but that substance is not *causing* it, only allowing it to become active. Novalis has this to say on the creative essence of Fire:

> The art of leaping beyond oneself is everywhere the highest act. It is life's point of origin, the genesis of life. The flame is nothing other than an action of this sort. Thus philosophy begins at the point where the philosopher philosophizes himself, that is to say, where he consumes and renews himself.[133]

Novalis calls light "the genius of the fire process" and, accordingly, Bachelard writes:

> Fire receives its real existence only at the conclusion of the process of becoming light, when, through the agonies of the flame, it has been freed of all its materiality.[134]

Light becomes the ultimate image of truth, which is the free, the undivided, the enlightened power of understanding arising as the "genius" of Fire's power of genesis.

It was previously stated that in living form the creative subsumes the mechanistic. Creation is "higher" than physical structure and process because it is the activity of the living idea (*entelechy*) which *creates* the structure and form of the organism. Already in the mobile imaginative thinking of the Water stage, we discovered (with Aristotle) that form is not merely structure but *the force for the realization of structure*. Water thinking means to participate cognitively in the plastic movement of those forces which mold physical substance into living form. With Air an opening is made in thinking, and physical organic structure and substance become transparent — they become a revelation of musical gesture (the toneless tone). Now it is apparent that the meaning of all formative process in nature (as opposed to mechanical process) *is* musical. Fire is the creative activity of mind which is commensurate with the originating creative impulse in living form, the self-active *entelechy* or "unmoved mover" (to use an Aristotelian expression).

Fire is the point where knowing becomes *an act of creation* — it becomes more like something we would call a cognitive *experience*. Production and cognition become one and the same; this unity, described by the philosopher Schelling as the essence of the artistic act, becomes heightened into a "scientific organ." The leap into a creative beholding is no leap into the unintelligible or the irrational, but it certainly reaches far beyond the intellectual mind. At this stage we speak of truth as creative as if we were leaping straight out of realism into idealism. Yet that leaping is precisely the nature of Fire, which liberates the creative energy hidden within the solidity of the physical form. This is "the energetic leap" spoken of by Goethe, who understood that cognition at this level must be thought of as creative deed; he writes: "[A]t this higher level we cannot know, but must act."[135] This is the "true theory" as Goethe understood it, a beholding which is paradoxically a doing. The philosopher Fichte conceived a remarkably apt expression when he spoke of "an activity into which an eye has been inserted."[136]

Life is the power which brings organisms into being and maintains their form up to the point of death, whence they disintegrate and return to "dust." When we perceive the organism with our physical eyes, we do not *see* this power but only the concrete results of its activity. The "eye of the sculptor" perceives the formative, molding activity of the *entelechy* — the "musical eye" apprehends its gesture. When we perceive the organism with the "eye of the spirit" (Fire thinking), we are actually experiencing this forming activity as a self-shining, self-generated spiritual truth, a creative impulse arising out of nothing but itself. Gesture is a manifestation of being; it is still apprehended from without, albeit through inner participation. The creative idea, the being itself, can only be grasped from the inside out. Kant spoke of the *intellectus archetypus*, the intuitive thinking which works from the whole to the parts, experiencing the wholeness (living idea or *entelechy*) that creates its own parts, that imparts itself. This is Fire thinking, the thinking that Steiner named Intuition. The Elemental journey of thinking is the progressive merging of one's thinking with the creative idea in nature, and Fire is the highest or most distilled stage. As Schelling declares: "To know Nature is, in effect, to re-create the world in one's own mind."[137]

An example of Fire thinking is Goethe's idea of the archetypal plant. When he first discovered this creative principle during his Italian journey, he wrote to a friend:

> I must tell you confidentially that I am very close to the secret of the creation of plants, and that it is the simplest thing one could imagine. The archetypal plant will be the strangest creation in the world, for which nature herself ought to envy me. With this model and the key to it, one can invent plants endlessly which must be consistent — plants that, even if they did not

exist, could exist — and not some artistic or poetic shadows and appearances, but possessing inner truth and inevitability. The same law can be applied to everything living.[138]

Goethe's description of the archetypal plant as "the simplest thing one could imagine" and "strange" requires further explanation. By "simple" he does not mean simplistic in the sense of easily explained or thought out. On the contrary, Goethe's "discovery" of the archetypal plant came as the fruit of a long and arduous quest, culminating in his Italian journey. The archetypal plant is not a complicated or sophisticated theory in the intellectual sense. It is "simple" in that it is the undifferentiated creative idea of the plant, its wholeness, which reveals itself in every part. That also makes it "strange" because it does not fit with the normal criteria for a scientific theory.

To think *in* nature, to be at one with the creative impulse of a living being — this is both the activity and the warmth aspects of Fire. Jochen Bockemühl writes of the significance of "warmth understanding" in the life sciences:

> We are intimately united with [warmth]. Warmth is an immediate presence, full of content but undifferentiated . . . The warmth enters us — our inner activity itself becomes an organ.[139]

Conventional scientific thinking might find it difficult to accept the idea that with Intuition we are *thinking in our will* — but this is the case. Our will is something which we normally consider to be more or less unconscious impulse, related to artistic expression and to productive activity in general, the very opposite of the bright, reflective quality of rational thought. Will is the *doing* that has its seat in the "lower human being," in the metabolic/ digestive and reproductive systems and the creative/productive movements of the limbs, which are all energized by metabolic warmth processes. The intense generative activity of this aspect of the human organization is polar to the largely unregenerating and immobile nature of the brain and nervous tissue.[140] Only when the eye of clear rational consciousness is "inserted" into our willing can it transform into the organ of cognition called Intuition. Thus we understand the movement from Earth to Fire thinking; the cognitive freedom of the Earth stage is not lost and is, indeed, what forms the will into an organ of cognition.

We have already seen how, with Imagination and Inspiration, two different aspects of human feeling take on a cognitive function — not particular personal feelings or emotions but *capacities* of the feeling life. Similarly, when we speak of will in connection with Intuition, we do not mean will impulses with a particular subjective intent or content. Consciousness enters into the *generative power* of the will, heightening its dynamic, flame-like nature. To "think in our will" means to awaken a human soul capacity commensurate with the creative impulse of the living idea in nature. Now we can complete our picture of the Elemental stages in terms of the soul constitution of the human being. Such is the aim of the pathway of the Elements — to engender a "thinking of the whole human being."

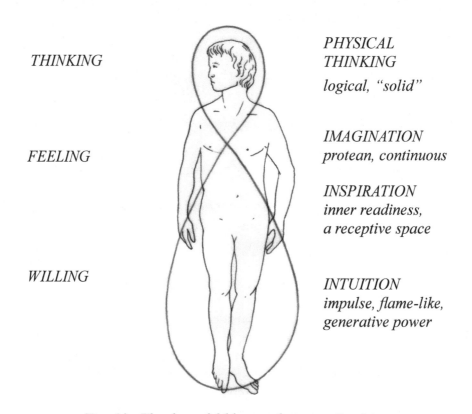

THINKING

FEELING

WILLING

PHYSICAL THINKING
logical, "solid"

IMAGINATION
protean, continuous

INSPIRATION
inner readiness,
a receptive space

INTUITION
impulse, flame-like,
generative power

Fig. 13. The three-fold human being — Intuition

Each stage of cognition necessitates a different form of languaging. At the level of physical cognition (Earth) we speak of conventional scientific discourse; this is the language of exact description and logical explanation, and, in a highly specialized form, it is the language of mathematical logic. The next stage, Imagination, calls for a plastic form of expression which is most obviously sculpture itself but includes painting, drawing and other visual arts. Speech becomes such a medium of expression when it realizes its potential to embody a sculptural thinking and exactly communicate the formative movements of the *entelechy*. The stage we have named Inspiration calls for an expressive medium or language imbued with musical sensibility. We have seen that this is not necessarily music as such, the sounded tone, but any language of gesture which reveals musical meaning (the example given was the four leafing activities). It is the need to find an authentic "language of creation" in the realm of the sciences that concerns us now.

Intuition, or Fire thinking, is that which the Aristotelians called *nous poietikos* — the active or creative thinking which thinks the creative process in nature. The term *poiesis* for the Greeks meant both creation (free origination) and "the word." Poetic speech thus had a fundamental significance for the Greeks, and they associated the poet-sage with the demiurgic powers or world-creating divinities.[141] The word was accorded primacy as the language of creation — not just any word but the poetic word, the originating word. For the pre-Socratic Heraclitus, the Element Fire stood for the *logos,* and *logos* meant both the deepest truth of the cosmos and the speech which gives utterance to this truth.[142] The poetical brings the true into the splendor of what Plato, in the *Phaedrus,* calls *to ekphanestaton*: that which shines forth most purely. Heidegger calls such poetic thinking the primal thinking on being, prior to the poetics of art but including it.[143] As he writes, "[t]he poetical thoroughly pervades every art . . .," and we could likewise say that it pervades the thinking of the whole human being, revealing the coming-into-being of living form, in both its sculptural and musical aspects.[144] "The poetical" does not necessarily mean poetry as such — poesy or verse — but more generally the living, poetical sensibility and the poetically infused word. This is the language of a Fire thinking, of genuine scientific Intuition.[145]

A "poetical science" was envisioned by the German poet Novalis, who shared many of Goethe's concerns regarding the methodology of science as it was practiced in their day.[146] Goethe, in his investigations into the science of living form, was aware to a high degree of the inadequacies of scientific discursive language to express the truths he had arrived at. An important part of his scientific work was his search for an adequate language, a search that arose out of his sense of frustration. He wrote:

> . . . thus the conflict between the perceived and the ideated still continues unresolved. We therefore take flight in the realm of poetry to attain some measure of satisfaction.[147]

Goethe worked with symbols to express his insights. Drawn from nature, and generally taking the form of verse, such symbols were for him "a living and instantaneous revelation of the inscrutable," able to tell us far more about nature than concepts.[148] In his *The Metamorphosis of Plants* he provides both a scientifically descriptive account and a poetic evocation of his discoveries of the creative laws of plant formation. Part of the poetic version runs as follows:

> From the seed it develops
> As soon as the Earth's quiet, fructifying womb
> Releases it into life:
> Into the quickening realm of the holy, ever-moving light
> Which nurtures the germinating leaves' first gentle
> unfolding.
> In the seed lay this force, simply sleeping:
> An initial, crude indication, self enclosed, curved under its husk,
> Leaf, root and germ, only half-formed, and uncolored,
> A kernel of unstirring life, held fast in dryness.
> Now, trusting in moisture, it sprouts,
> Lifts itself out of the darkness.
> Simple in shape, like all children. . .

And a little further on:

> You are filled with ever-fresh wonder
> When flowers unfold on their stalks
> Above the leaves' multiform framework.
> In their glory, however, they herald a whole new creation.
> Yes, those pure-colored petals feel the touch of God's hand,
> And, quickly contracting, they throng,
> Twofold in circle and center,
> Destined for union they stand,
> Glowing in gold and in sweetness,
> Festive as pairs round the altar. . .[149]

It is the poetic account which brings forward the mighty idea that expresses itself in every plant, even a humble herb of the field such as hedge mustard. The growth and metamorphosis of the plant express a drama of universal significance; it is a drama of the polar forces of light and darkness working one upon the other, forces that live in tension and resolution, that develop and unfold, that speak their meaning in a "language of creation."

Thus we come to the essential Intuition of plant form: in the plant, Earth realizes itself as Sun and, equally, Sun realizes itself as Earth.

The seed sleeps in the Earth. Earth and seed belong together in their sleeping seed-natures; both are rounded, relatively undifferentiated, but receptive and imbued with the potential for form. In the simple rounded shapes of seed and seed-leaf we perceive the childhood of form. In the flower the form, the idea, has fully awakened, is made manifest as shape, number, scent, color. Flower and Sun belong together as the sphere of the luminous, the awakened. The idea is active, formative. Working from the plant's future, it progressively gathers Earth substance unto itself, distilling it, heightening it. This is the creative activity of the plant. The transformation of Earth substance allows the idea to become manifest. Inert, dark, inactive Earth substance is subsumed and sublimated, is given over to the impulse of form. It is the same with works of art: physical substance — pigment, stone, clay — is formed and transformed so that it becomes transparent to the *idea*. When we see a work of art we do not see substance as substance; we see the manifesting idea. This is the artistic meaning of the plant — substance is spiritualized, substance becomes the vehicle, the expression, of the creative idea. Intuitively understood, the plant is nothing other than a work of art.

The morphological language of the plant reveals a union of spirit and matter. The structure and function of the flower is itself an image of this marriage. The green, Earth-like ovary below, containing the seed-like ovules in its moist dark interior, is enveloped by the luminous and hollow "Sun-space" of the petals and ringed around by the golden stamens.[150] The penetration from above of the sperm into the ovary and egg and the union of the egg and sperm to form the zygote is the enactment of this union. This process needs to be understood inspiratively and intuitively: in essence, an active Sun process seeks to penetrate and merge with a receptive Earth process. Only through such a deeper understanding can we arrive at any kind of living insight into what happens at the chemical, chromosomal level. It is quite unnecessary — and actually misleading — to speak of male and female organs, or even of the sexual process, at the level of the plant. Goethe himself struggled with this question, although he felt obliged to refer to "male" and "female" parts as dictated by convention. However, he also writes in his essay *The Metamorphosis of Plants*: "...we are inclined to say that the union of the two genders is anastomosis on a spiritual level ..."[151]

Looking now specifically at the Cruciferae family: the idea of these plants is strongly formative and active. It calls the substance aspect of the plant toward itself, molding it, intensifying it. This idea is a living musical interweaving of forces, a unique patterning of gestures which constitutes its familial and species natures. The familial nature of this plant manifests especially in the characteristic "cross" of its floral form — symmetrical, static, in balanced simplicity or calm repose (in comparison with the more dynamic five-fold form, characteristic of many plant families). This oppositional, static quality is carried into the two pairs of opposing stamens and the superior ovary — a two-locule opposing form with a simple style. The calm repose of the flower is carried aloft in a way also characteristic of the family. Alternating leaves bring an essential dynamic, a vitality of gesture, along the axis of the plant, and this is amplified in the alternation of the racemose floral form. In this manner the floral raceme manifests the presence of vegetative growth forces in the realm of the blossom.

Fig. 14. Hedge mustard — alternation pattern leading to symmetry (gesture sketch)

So, firstly, we take the music of the overall plant structure of this family and intensify it intuitively: an alternating, dynamic gestural structure consummates itself in gestures of symmetrical repose or "still-points." Something of this gestural progression can be heard in the classical *tierce de Picardie*, where the complex "soul movements" of a piece in the minor key terminate in the simplicity of a major chord. Here there is not just a concluding cadence (and every flower has something of a cadence in it) but also a terminal change of mode.

Fig. 15. Tierce di Picardie (from J.S. Bach, Invention in E minor)

In the previous section we looked at the specific metamorphosis of hedge mustard, at the way its lobed, symmetrically divided leaves contract and "dematerialize," giving rise to the symmetrical stasis of the flowers. At the same time as *pointing* refines these leaves to arrow forms high on the plant axis, held close to the stem, the stem is elongating between nodes and stretching upwards. The flowers arise at the extremity of this stretching. We do not sense the flowers of this plant as resting on the Earth or as belonging in any way to the realm of darkness — and this is true of the other crucifers, even when this stretching-elongation is not so accentuated. The simple symmetry of their flower form is attained in the heights, relative to the form and upright habit of the whole plant. The flower is held aloft from the vegetative organization — it belongs exclusively to the Sun-sphere. This is something we read in other features as well: we see it in the characteristic superior ovary (it is free and "has left leaf-nature behind") and also in the gestures of the principal colors of the family.[152] These are the "active" colors, closest to pure light — white, yellow and orange — as opposed to the "passive" colors of blue-violet, and even red, which are all "overcast with some darkness."[153]

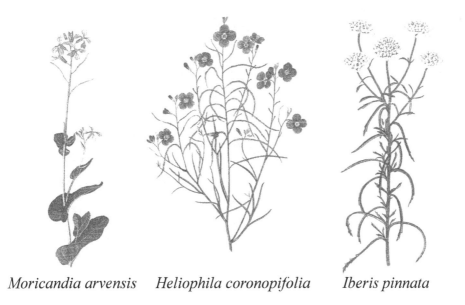

Moricandia arvensis   Heliophila coronopifolia   Iberis pinnata

Fig. 16. Comparison of crucifers (after V. Heywood, Flowering Plants of the World [London: Batsford, 1993])

Now we can marry these preliminary Intuitions with insights deriving from other aspects of the plants' nature. It has already been noted that hedge mustard and most other crucifers express an enormous vigor and vitality in their growth habit, even in relatively dead and desiccated habitats. Pelikan calls them "soldiers of the sun," wresting life from even the poorest soils.[154] They grow rapidly, dynamically elongating and stretching upwards to realize their static "Sun-points." This inherent vitality does not rigidify into an earthy hardness in the vegetative plant and neither is it turned — "in the heights" – into luxuriant, highly scented and succulently formed flowers, nor to juicy and delicious fruits. Their flower forms are tiny and pure in their static simplicity and their characteristic whites or shades of yellow/orange. The plants hold back an outward expression of their immense Sun-vitality in the form of their flowers and instead focus it into a *substance*-expression — seed substance in some species (giving oils and mustard condiments), vegetative substance in others (in many species giving fodder and food crops). All these plants also express their Sun-vitality, to some degree, in the form of a fiery sulphurous pungency. In the case of the genus *Brassica*, humans have been able to magnify the natural tendency of the plant and hold back this abundance of vitality, "damming it up" in different organs of the plant, which develop considerable vegetative substance. The result, as discussed in the Water stage, is the different varieties of cabbage. Somehow, ancient horticulturists were able to "invent" — out of their intuitive grasp of the creative principle inherent in this plant family — new species of Cruciferae, "that, even if they did not exist, could exist..."

Thus we come to the end of our journey into the nature of living plant form. In a larger sense, it has been a journey toward a science of the wholeness of nature, an authentic science of living form, which develops from a thinking in which the whole human being has become active. In Goethe's way of putting it, a true life science must spring from "all [the human being's] united powers." Moving along the Elemental pathway from Earth to Fire, we have engaged the gamut of our cognitive capacities. In doing so, we have united the factual with the creative, thus realizing the aims of Goethe's scientific method as stated in the previous chapter: "[P]roductive power of the imagination combined with all possible reality."

PART 2

A GOETHEAN STUDY OF PLACE

I shall endeavor to find out how nature's forces act upon one another, and in what manner the geographic environment exerts its influences on animals and plants. In short, I must find out about the harmony of nature.

—Alexander von Humboldt[155]

Is not every flower a type of flame?

—Gaston Bachelard[156]

A musical thought is the one spoken by a mind that has penetrated into the inmost heart of the thing; detected the inmost mystery of it, namely, the melody that lies in its soul, whereby it exists and has a right to be here in this world. All inmost things, we may say, are melodious; naturally utter themselves in Song ... See deep enough, and you see musically; the heart of Nature being everywhere music, if you can only reach it.

—Thomas Carlyle[157]

CHAPTER 3

EVOLUTION, PLACE, AND THE ORGANS

OF LANDSCAPE

Introduction

To properly understand what is meant by the Elemental modes of cognition one must literally *do* them. Together they represent a pathway, a methodology for organic thinking, and to treat them theoretically or merely philosophically is to miss their relevance to modern scientific and artistic praxis. Goethe's participatory way of science is that of the engaged "nature observer," the researcher who thinks in and with the living forms and processes of nature, not merely *about* them. The ideas on the Elemental modes developed up to this point are intended as a preparation for the ecological research which is documented in the following chapter.

Goethe had an influence on the development of modern ecology, in an indirect way, through his early association with the biologist and geographer Alexander von Humboldt. What Humboldt and Goethe had in common was an artistic sense of nature which they both felt was indispensable to the development of life science. In his study of broad geographical environments, Humboldt was closely related to Goethe's aims in his morphological studies — what both sought was insight into the wholeness or what Humboldt called the "harmony" of natural organization. Humboldt was neither a poet nor a musician, but he had a highly developed aesthetic sense from a lifelong study of the visual arts. His way of conceiving nature as a harmony or unity suggests a degree of musical insight, a perception that the natural order is essentially a *musical* order. Goethe, too, sought the wholeness of organic form, and it was his strongly sculptural sensibility that led him to his appreciation of metamorphic processes in nature.

The artistic aspect of Humboldt's work did not advance through the way his methods were married with the established biological sciences. What actually came into being was the modern science of ecology, the quantitative study of ecosystems, which is underpinned by Darwin's hypothesis relating to evolutionary development. Humboldt's organic world-conception became tied to the Darwinian mechanistic view of evolution which asserts that transformation comes about through causal interactions between different aspects of geographical environments — in other words, by the hypothetical mechanism of natural selection. Today we may open the way toward a Goethean science of landscape by taking cognizance of the original aesthetic inspiration of Humboldt's ecology and drawing it back into the context of a Goethean artistic methodology. We do this in recognition of the fact that a landscape is a living entity, requiring an appropriate "living" methodology in order to be properly understood.

Conventional ecology, deriving like the other sciences from a theoretical or "solid" form of thinking, takes an external objective view of nature whereby the human "I" experiences itself as an onlooker on nature's processes. Yet this is at odds with the original intuition of ecology (*oikologie*) as the "household" of nature in which the human being exists together with the other beings of nature, not just as an observer but as a participant in the natural order. Living thinking is the thinking of the whole human being, and we have already seen how, in the thinking that seeks to understand living form, both the scientific and the artistic in human nature must become active. This means that — in a way which seems paradoxical to the purely logical mind — the environmental researcher becomes both a knower and an experiencer of landscape. A true ecological science must include the cognitive experience of the researcher, and the work — the ecological study itself — must be realized as a cognitive-creative deed, consciously developed as a dimension of the landscape.

Much has come to pass in the modern world as a consequence of the externality of our conventional ecological perspective. The normal scientific outlook sees human productivity as a process of appropriating the mechanical laws of nature and using these for transforming or otherwise managing nature based on the model of a machine. Human beings are not just knowers but are makers or creators; they have radically transformed nature in the process of creating human culture, but this culture is ever more at odds with the very nature from which it springs. But if the human knower were able to enter into and understand the creative forces working in nature, and then extend and develop these in the formation of human culture, then the relationship of nature and culture would reach a new level of mutuality. Such is the aim of Goethean science in relation to ecological education and research; it seeks to articulate a methodology which will open the way toward a holistic wisdom in our understanding and transforming of nature.

The aim of the particular research set forth in this section is to present a Goethean approach to the study of landscape such as could be taught at the college or university level. This approach could also find application in other cultural activities ranging from architectural and landscape design, to medicine, to agriculture. The Goethean science of living form will gradually reveal its potential to act as a bridge that unites creative process in nature with human creative endeavor. In this way it will help us develop a social ecology in the fullest sense.

Evolution as Creative Process

We proceed from a Goethean understanding of plant form to a broader picturing of the lawful structure of ecosystems, and to a living understanding of evolution. Although Goethe's writings deal mainly with aspects of plant and animal morphology, these writings are actually expressions of a view of the evolutionary process which he shared with various other notable thinkers of his time, in particular Herder, who was an early associate. Such ideas were further developed in the natural philosophies of Oken and Hegel. To the extent that this stream of German thinking was able to perceive nature with the "eye of the artist," it arrived at some extraordinary insights into evolution as a creative process, insights which have been largely misunderstood in subsequent years.

To grasp Goethe's artistic view of evolution, it is necessary to have understood his approach to the study of organisms. We have seen that his understanding of organic form was inspired by the Aristotelian tradition; specifically, he saw *entelechy* as the creative, *forming* aspect of organic form (or the force for the realization of form) rather than *that which is formed*. One of Goethe's greatest inspirations from the Aristotelian stream was the philosopher Giordano Bruno, who spoke of the "inner artist" of nature, of nature creating itself out of itself, of nature as living creative being.[158] By contrast, the central theological outlook of European culture was related to the Platonic view, which conceived of all living forms as imperfect copies of eternal and perfect Ideas or archetypes that exist "in the mind of God." This world of divine Ideas was considered to be a metaphysical world, entirely beyond mundane existence yet influencing it. According to this view, each species was created with a preordained purpose on the face of the Earth; the notion of evolution, the idea that an organism could be modified as a result of earthly influences, has no place here.[159]

Darwin overcame the problem of the immutability of the archetype in a way which was in accord with the materialist outlook of his time. He took the metaphysical notion of the archetype and made it something entirely material — namely, the progenitor organism from which present organisms evolved by the process he called natural selection. He claimed that a "selective pressure" causes one organism to survive, another to die out. Neo-Darwinist ecology seeks to measure and analyze these evolutionary processes and to explain them, which it does increasingly through the modern science of genetics. Thus the Darwinian approach dispensed with the traditional teleological conception of divine causation, the view that each organism is created with a fixed pre-ordained purpose (*telos*). The brave new neo-Darwinian outlook is that organisms derive from more primitive organisms through chance mutations in those primitive organisms; these mutations, it is thought, *just happen* to result in organisms that are better adapted to environmental conditions.

The analytical science of ecology, working within the neo-Darwinist frame of reference, sets about studying the interconnections between the minerals, plants and animals which make up a natural environment. From the point of view of analytical ecology, what we have before us in a landscape is an enormous diversity of forms which are linked through causal connections — soil type causing the appearance of certain types of vegetation, vegetation types causing the appearance of certain animal types, and so on. But no number of such studies will ever add up to an understanding of the formative potential of this landscape as a whole. Goethe, on the other hand, sought the archetype as it works in plant and animal form, responding creatively to changing environmental conditions. As Rudolf Steiner explains:

> For Goethe, the individual changes [in plants and animals] are the various expressions of the archetypal organism that has within itself the ability to take on manifold shapes and that, in any given case, takes on the shape most suited to the surrounding conditions in the outer world. These outer conditions merely bring it about that the inner formative forces come to manifestation in a particular way.[160]

The archetypal organism is another name for the wholeness or creative living essence of the organism which Goethe saw as being engaged in a kind of conversation with the external conditions of a landscape. A great deal here depends upon a proper understanding of what Goethe meant by archetype. As we have seen in our exploration of the Elemental modes of cognition, the archetype is not a physical entity, but neither is it merely an abstraction of the intellectual mind. Rather, it is something actually *experienceable* through imaginative intelligence. Every plant which makes up the plant kingdom is a specific expression of the same archetypal plant; the archetype is unchanging, yet it actualizes itself over time in the multitude of plant forms and, in that sense, is in a constant state of transformation.

Working imaginatively, inspiratively and intuitively, it becomes possible to experience a landscape creatively in the same way that Goethe grasped the creative processes at work in particular organisms. The picture emerges of a landscape as a self-creating wholeness, coming into being in the way a plant comes into being, each element an expression of the wholeness of its living form. As Goethean biologist Mark Riegner writes:

> [T]he wholeness of a place comes to expression *expansively* in the overall landscape and *focally* in the parts of the landscape.[161]

What this amounts to is a revolution in ecological understanding. The wholeness of a landscape comes to presence in any particular animal or plant entity just as it does in the total geographical context of the landscape. The wholeness, in other words, is "written" in the morphological language of every aspect of the landscape, and this is so because the wholeness, the archetypal landscape, is the source of the parts, because it has "imparted" itself as the form of the landscape.

The holistic understanding of landscape thus requires the development of the capacity to perceive evolution as growth and creative process. We noted in Chapter 2 that a thinking which only seeks explanation is inadequate at the creative level. The creative dimension of evolution must be understood creatively — through imaginative thinking. The effort to explain the existence and form of living things has led to some curious, even preposterous, notions. It is said that bacteria evolved into plants, animals and humans, meaning that bacteria actually *turned into* humans — very slowly, of course. It is even said that non-life turned into life, that a mixture of water and other inorganic chemicals was so stimulated that it eventually took shape as a fully-fledged organism. Such ideas are stated as facts in most biology textbooks today, sprung as they have, inevitably, from the scientific world-view that sees organisms essentially as mechanisms.[162]

A living, creative understanding of evolution develops in exactly the same way as does a living understanding of plant form. Imaginative thinking enters into and participates in the changing forms of the plant organs and thereby grasps the formative idea (*entelechy*) of the plant organism. Such thinking is able to perceive a formative lawfulness in the "leaps" — for example, between lower and upper leaf, between leaf and flower. Exact sensory imagination unites the changing forms of the plant in a single generative movement but feels no requirement to explain the process in terms of one leaf physically *turning into* another. It is the same with evolutionary growth; imaginative thinking enters into and participates in the changing forms of nature over time and is able to grasp evolutionary processes as fluid formative movement without any requirement to prove that one organism physically *turned into* another.[163] We can learn to understand evolution creatively by training our imaginative organs of perception through a Goethean study of metamorphosis in plants.

The Landscape and its Organs

Natural undisturbed places in nature (but also certain places transformed by human beings) have a sense of uniqueness, a palpable quality of harmony between their various parts, and this quality is what in ancient times led people to speak of a *genius loci* or "spirit of place." The word "genius" relates to "genesis," to the living creative spirit of a place, not sense-perceptible yet expressing itself in all aspects of the place. We might, using the expression of Bruno, call it the "inner artist" of the place, its formative potency. The *genius loci* was grasped intuitively or even instinctively, not with the clarity and exactitude of conceptual thinking, and so has commonly been considered akin to a mystical experience and nothing more. It is nevertheless true to say that the notion of "spirit of place" has been most influential down the ages; it has inspired landscape painters, architects, farmers, city-builders and others who sought guidance in how places were to be transformed for human purposes.

This idea of landscape as living being is not something that is generally embraced by the conventional scientific way of thinking. Ecology speaks of an ecosystem as the web of relationships between living and non-living parts but not of the totality of a landscape as a living entity. Yet, as the Goethean biologist Andreas Suchantke writes, "[f]rom landscape as "ecosystem" it is only a short step to landscape as "organism.""[164] Basing his research on in-depth phenomenological study and on insights of Rudolf Steiner into the three-fold nature of living organization, Suchantke explains that most of the distinguishing features of an organism also apply to ecosystems. The ecosystems he refers to vary vastly in scale; they range from a pond or lake to a whole continent, and, in connection with the latter, he makes a significant point. He suggests that the geographical elements which make up a continent — for example its climatic zones, its forests, mountains and deserts — should be considered as organ-like structures of the greater whole which is the Earth. It is upon this basis that he proceeds with his research into the landscapes of Africa and New Zealand.[165]

In fact, the idea that landscapes have a living organization — that is, an organization of organ-like structures — had already been put forward by thinkers who had a connection with Goethe's way of science. Alexander von Humboldt considered every part of the Earth to be a reflection of the harmonious unity of the *Weltorganismus,* of the cosmos as a living whole. His close associate, the geographer Carl Ritter, spoke of the continents as "organs" and the parts of the continents as "individuals," each of these being members of the "organism" which is the whole Earth. The basis for this way of thinking was derived from actual observation of geographical phenomena. Objects of botanical study are divided according to kind, but geographical phenomena are meaningful mainly in terms of their spatial connection to the Earth's surface as a whole. Ritter's method was not to gather up a mountain of information about a particular landscape but rather to seek out the essential character expressed through its interrelated principal features, both animate and inanimate. This irreducible "essential character" we might also call the "wholeness of the landscape," its living, formative essence.[166]

The natural philosophy of Hegel also explores the living organization of the Earth as a whole.[167] His research helps us to comprehend the different levels of organization in landscapes and thereby to come to the *cognition* of landscapes as living entities. Hegel was seeking to raise primal intuitions of landscape as living being — intuitions which were alive in traditional European spiritual and artistic culture — to the stage of what he called "developed intuition," to actual knowledge of the living organization of landscape.

The Geological (or Geo-solar) Organism

Hegel speaks of the *geological organism,* not as actually alive but neither as dead in the sense of a mechanism. He describes the geological organism as the basis of all earthly life and as showing an incipient life in every respect. Hegel alludes to different "themes" in the mineral realm which are recapitulated (although transformed) in the plant and animal realms; for example, he notes that the "crystal theme," the straight lines and planes of crystal forms, can be observed in the lines and planes of plant forms, less so in the rounded forms of animals (but still evident in the linear formations of skeletons). Such a study could go very far indeed; we may, for example, look at the inherent tendency of water to form waves and vortices and how this prefigures all manner of rhythmic form and function in plants and animals. We see it in the ubiquitous vortical formations — from sea shells to the semi-circular canals of the human ear. Steiner was later to speak of "prophetic forms," meaning those forms which, while more primitive on the time-line of evolution, "look ahead" and in a sense predict the forms that later emerge.[168]

The principal "organs" of the geological organism — Hegel calls them its "members" — are the rocky mineral body of the Earth, its waters collectively, its atmosphere (and all gaseous formations) and its warmth or fire nature. These are the four discrete dimensions of the physical realm which work continuously on and through each other: water on rock, air surrounding and entering water and soil, warmth permeating all the other elements and causing movements on scales vast and small. Mountain building, tectonic plate shifting, volcanism, metamorphosis and erosion, weather formation, chemical reaction — all these complicated processes on a planetary and local scale relate to the interactions of the four principal "organs" of the geological organism, and every landscape is a unique localized expression of these interactions.

Only an exact imaginative thinking that can participate cognitively in these processes, a thinking which gradually builds a fluid imaginative picture of the whole interweaving of these "organs" on a planetary scale, can begin to penetrate to the unity dimension of the geological organism, its living organization. Goethe himself came to realize that any particular rock structure or geological process is meaningless when viewed in isolation; he saw that they must be understood within a regional context, and this in the context of the whole Earth. What for the plant and animal is the integrity of a "body" is for the geological realm the Earth as a whole.[169]

Goethe only progressed a certain way in his contemplation of the geological realm and, indeed, stopped short of a full understanding of the geological body. Cognitive imagination leads us to the discovery that the Earth, the planetary entity which lends the prefix "geo-" to this field of study, is but one side of a polarity — with the other pole being the Sun. To this extent Hegel's term "geological organism" is also somewhat misleading. Earth and Sun, in their rhythmic interweaving through the days and the seasons, represent a unity. It is true that modern physical science has moved some way toward perceiving this unity through its grouping within the family of the geosciences the traditionally independent studies of geology, meteorology, hydrology and solar physics. Actually these are not just "geo" sciences at all, for this family includes sciences which pertain to the extra-terrestrial nature of the Sun. The proper name for the being which we are endeavoring to understand is the "geo-solar organism" or, better, "Earth—Sun" (which is part of larger organizations — the solar system, the Milky Way and so on).

The massive mineral body of the Earth exerts a pull toward its center, this *centric* force influencing every entity on its face. Every landscape bears witness to gravitational forces and processes: the weight of every stone which rests on the Earth's surface, the processes of weathering and sedimentation, the hydrological cycle, precipitation and the seaward flow of rivers, pressure metamorphosis and so on — all such activities involve cause-and-effect relations that may be understood primarily through the action of gravitational, mechanical or more generally centric forces.

But gravitational force does not bear upon the warmth processes which also work upon massive bodies in every landscape. Warmth is ever tending away from the Earth; it is a manifestation of universal levity or lightness which has an immediate focus (with respect to our solar system) in the Sun. The evaporation of water, the uplifting movement of convection currents in liquefied rocks of the Earth's mantle and in the oceans and atmosphere — all these thermal processes of the geo-solar organization display the working of levity forces, tending away from the Earth's center. We also see levity forces at work in the lightness or buoyancy of water and in the expansive tendency of the atmosphere.

Gravity and levity are polar forces that have their immediate focuses in Earth and Sun. Those working with Goethe's scientific methods have gone a long way toward determining the nature of this geo-solar organization by showing the one-sidedness of a physics (and, as a consequence, a geology, hydrology, meteorology and solar physics) which seeks to explain everything in terms of gravity and mechanical/causal interactions — inspired, we might say, by the falling apple rather than the rising seedling.

In the domain of the geo-solar sciences, Goethe's way of science faces the same kind of challenge as it does in color theory. For Goethe it was immediately evident from observation that darkness is an active force in its own right, a force which, on meeting the activity of light, gives birth to the colors. He insisted that a physics that conceives of light alone as the reality is a one-sided physics. It is the same with our approach to the geo-solar organization: the phenomenon of levity (lightness) is not merely the diminishment or absence of gravity but is an actual force in its own right. Much can be learned from the advanced polemics of "Goethe contra Newton" in the area of color theory and applied to the newer fields of terrestrial and solar physics.

The result of the universal application of a Newtonian style of physics within the physical sciences is that Earth and Sun have been approached as if they are exactly the same kind of entity — that is, massive bodies. Thus we are brought to a consideration of the essential nature of the Sun, although the subject can only be touched upon here. Solar physicists are currently endeavoring to demonstrate that the Sun's center is immensely massive and weighty in line with conclusions arrived at long ago through Newtonian celestial mechanics, which provided an explanation for the orbits of the planets. This approach, all along, has been unphenomenological and thoroughly theoretical in character; in the process, assumptions have been made that are currently coming into conflict with discoveries which can only be incorporated into this theory with immense difficulty. In the last ten years, through the rapid advance in the technologies necessary to view and analyze the Sun, longstanding assumptions about the nature of the Sun have had to be thrown aside.[170]

The first and almost facile observation which can be made is that the Sun is a being of light and warmth; everything that we know about the Sun as an immediate perceptual phenomenon pertains to its *lightness* (implying both luminosity and levity). This is where the Goethean phenomenology of the Sun begins and from where it draws its eventual conclusions. That these conclusions directly contradict the conventional mechanistic/quantitative view of the nature of the Sun as a massive body (which, after all, is a hypothesis yet to be verified) certainly makes them appear radical, but that in itself is not a reason for them to be ignored. In the briefest terms these conclusions are: the Sun (or any star) is not a gravitational center at all but rather a focus of universal levity forces. The application of modern synthetic projective geometry to the understanding of these forces has placed the question of the nature of the Sun on a sound footing in mathematics, showing the character of the space represented by the Sun to be of an entirely different and opposite *quality* to that of the Earth.[171]

Working empirically with the phenomena of Earth and Sun toward inspirative and intuitive perception, we arrive at the primary gestures of gravity and levity as characterizations of the dynamic *qualities* of these polar entities. As with all polarities in nature, gravity-levity has an archetypal quality — it is an "archetypal phenomenon," to use Goethe's expression. Gravity-levity is the primary formative idea of all earthly landscapes and could therefore be called the "archetypal landscape."

The living idea of the geo-solar organism can now be graphically indicated in a lemniscate form with a three-fold structure, representing the polarity of Earth-Sun (or gravity-levity) united through rhythmic seasonal and day/night interweaving.

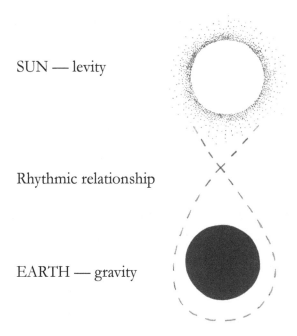

SUN — levity

Rhythmic relationship

EARTH — gravity

Fig. 17. The geo-solar organism (gravity-levity)

The plant organism

Another essential aspect of landscape is its plants. Even the most arid desert and frozen regions of the Earth present some form of plant life. Goethe sought to uncover the nature of the "archetypal plant," the creative idea which is the source of all particular plants. The science of ecology only developed after Goethe's time, and he did not arrive at the point of being able to show how particular plant species are expressions of a landscape as a whole. That is a task left to those who are developing his way of science within the modern discipline of ecology.[172]

The plant has a unity of shape; all its members are physically linked in one coherent form or body. However, the members or organs of a plant entity are related quite differently from those of an animal, which are enclosed within a unifying skin. The plant organs — root, leaf, flower and fruit — are apart from or external to each other even if connected.[173] What distinguishes the plant organization from the geo-solar organization is that the plant organs *come forth from each other;* that is, the plant *grows*. Its organs exist as potentialities in the seed and come forth sequentially in rhythmical fashion through the life-cycle of the plant.

Let us suppose that we have worked at length to understand the geo-solar organization through the practice of exact sensory imagination. This requires imaginatively "flowing" between the polarities of Earth and Sun, moving through the rhythmically interacting physical processes of earth, water, air and warmth, thereby forming a dynamic inner picture of this organization in its totality. What gradually emerges from such imaginative work is the living idea of the *plant*. One can inspiratively and intuitively "see" that the plant is the image or embodiment of the rhythmical interaction of Earth and Sun. To put it another way, the plant is a further expression of the living idea that can be "read" in the nature of the geo-solar organization.

Qualitatively, the roots are unified with the whole mineral body of the Earth and their downward orientation expresses the gravity principle. The roots are characterized by fixity and hardness (in a tree these qualities carry upwards in the form of the sclerophyllous trunk).[174] The taste of the root is generally bitter with a mineral saltiness, and this bitter taste has a contractive gesture which is akin to the in- or downward-pulling gesture of gravity.[175] Roots have a linear tendency, radiating out from the central point of the seed, which itself is of a hard, mineral nature. They express a centric (point-centered) Euclidean geometry and are perfectly adapted to the space-filling character and rigid geometry of mineral substance. Overall, in their immobility and earth-rootedness, plants reveal their affinity to the static, crystalline formations of the Earth.

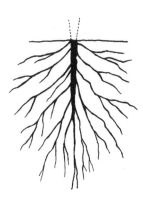

Fig. 18. Roots expressing centric (Euclidean) geometry

Moving upward to the blossom we find quite polar qualities expressed. Flowers generally grow toward the sunlight, this being the most obvious expression of their levity character. But the cognitive imagination can discover much else that relates the flower to light and warmth. We may firstly consider its colors, which are truly "born of the light."[176] The levity character is also evident in the exuded ethereal perfume and sweet substances developed in the nectaries and later more dominantly in the fruit. To the cognitive imagination the sweetness of taste and perfume have an expanding, uplifting gesture, polar to the roots' bitterness/contraction. The coloration, sweetness and fragrance of flowers express — in shades and degrees particular to each species — the dynamic qualities of "lightness" and warmth. It is for this reason that Gaston Bachelard speaks of "the *unity of fire* between sun, tree and flower."[177]

The linear roots fill the space of mineral substance with a network or grid. However, the floral space has an opening, embracing gesture foreign to the mineral realm. This is a hollow, cup-like space, which can vary from the tubular to the very opened-out. It is as if the roots' "Earth-space" were turned inside-out, inverted, made receptive and fecund. The focus of this "Sun-space" is the stigma, which the insect touches with a grain of pollen; but the geometry of this space is not centric and metric. Goethean biologists have shown the capacity of modern projective synthetic geometry to describe precisely and evoke exact imaginations of the quality of this space.[178] The floral space is a "counter-Euclidean" space, not governed by a point-center but by an "infinitude within," which is an infinitely receptive center created by living form. It is through the development of such spaces that the plant achieves an intensity of lightness or levity, polar to its mineral-related gravity.

Fig. 19. The flower expressing a hollowed "counter-Euclidean" space

Between the polarities of root and flower is the vegetative plant. Here the cognitive imagination finds the predominating character of *rhythm* expressed in the periodic appearance of the leaves and (in many plants) the spiraling phyllotaxis. This rhythm is the concrete image (or symbol) of the rhythmical relationship of Earth and Sun. From each growing point the leaf pairs or whorls form an enclosing space before flattening onto the horizontal. In this way the leaves "speak" in advance of the levity-space which only fully establishes itself in the form of the flower. The vegetative plant is the "in-between," mediating and relating root and flower, the polarities of gravity and levity, transforming the one into the other.

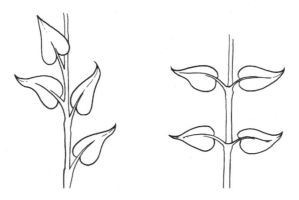

Fig. 20. The vegetative organization as the rhythmic "in-between" of root and flower

Implicit in this process of transformation or metamorphosis of the plant is what Goethe called intensification, enhancement, or heightening (*Steigerung*). Metamorphosis takes place in a living form when a principle that is concealed in the early stages of the growth process (and is only visible to the "eye of the imagination") is progressively revealed and elaborated in the later. As Rudolf Steiner expresses it, in the phenomena in which enhancement reaches its goal, "the idea becomes immediate truth and becomes ... visible to the physical eye."[179] Thus we can say that the flower stands at the summit of the vegetative development of the plant; the flower is the revelation of an idea that is more or less hidden in the form of the leaves. This is something we have already explored in the previous chapter, in the discussion of Air and Fire thinking in connection with hedge mustard.

We can now represent the flowering plant in a similar way to that of the geo-solar organization, as the expression of a three-fold principle. The polarities of root and flower are images of Earth and Sun (gravity-levity); these are united through the rhythmic formation of the vegetative part of the plant.

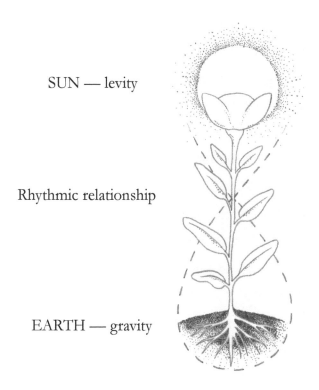

SUN — levity

Rhythmic relationship

EARTH — gravity

Fig. 21. The flowering plant as the image of the archetypal landscape
("gravity-levity")

The animal organism

The next organ of the landscape is its animal life; as with plants, some form of animal life is expressed in even the most barren terrestrial landscapes. We have already noted that the structure of the animal body is quite distinct from that of the plant. Taking a mammal as an example: all its organs are contained within a single unifying organ, a skin. Each organ of the plant — the root, the leaf, the flower and fruit — is still something of an "individual," each exists separately from the others even though they are all parts of the same plant. By contrast, the organs of the animal are "gathered" through the power of an individuality of a higher order than the plant, and the skin is an outward expression of this individuality (of course, there are differences in this regard between "higher" and "lower" animal forms).

"In the animal, light has found itself," declares Hegel, and this statement — at first glance utterly abstruse — is intelligible in terms of the metamorphosis we are considering here.[180] Hegel calls the animal an "aethereal being," a being of warmth and light. The archetypal landscape is manifested in a very definite way through its animals: in animals, light and warmth have become inner life. The "inner flowering" of animality, or the animal soul, expresses itself as different forms of desire. Bachelard seeks an authentic Imagination for the inner being of the animal and finds it in the words of Novalis: "Flames constitute the very being of animal life . . . the flame is in some way a naked animality, a kind of excessive animal."[181] We could follow this by asking in the manner of Bachelard: is not the soul-nature of every animal a type of flame? The animal soul as grasped through imaginative-inspirative cognition is the metamorphosed Sun, Sun become inwardness or sentient life. The plant, through its growth process, reaches *toward* the Sun, and its highest point, the flower, remains only ever an image of the Sun. Each flower presents what can be called an objective soul quality — a mood or atmosphere which takes on concrete form, color and scent. This is why Hegel calls the flower the plant's "highest subjectivity."[182] With the animal, mood or atmosphere has become an inner life of feeling, actual subjectivity (although not ego). Cognitive imagination comprehends animality as a further realization of the living idea which is the archetypal landscape. The polar gestures of Earth and Sun, still held apart in the plant as root and flower, unite and enter into each other, transforming each other and giving rise to an altogether new kind of being — the animal.

Soul is not sense-perceptible because it is not a physical substance; it is not possible to locate it in space and therefore it is not possible to quantify it. This means the soul is not amenable to analytical science — but it does not follow from this that soul cannot be exactly understood. The supersensible form of the soul appears *through* the sensible physical form of the animal which it animates. Soul is the creative element in animal life because it is soul which imparts itself in the overall form and inner organization of the animal as well as in the form of its behavior. Soul is the "wholeness" of animality, the animal *entelechy*. To study the animal analytically is to bring to light the functional relations between its organs — for example, the movements and effects of hormones on the different organs. Never, however, will analysis grasp the nature of the soul because such thinking does not attain to a creative stage. The Goethean phenomenology of animal form, proceeding from Earth to Fire thinking, allows the sensible form of the animal to become transparent to its fiery, creative soul principle.

A true zoology must reach the level of soul because this is the distinguishing feature of animality. Animals, through outward expressions of their inwardness, determine their lives in a manner quite different from that of plants. The plant is embedded in nature, in the maternal Earth with its liquidity and in the light-filled atmosphere. It is, as it were, entirely asleep, and in this sense it remains close to mineral phenomena, which are "moved" by the world and not self-determining. A single plant is scarcely individualized out of the elemental realm — as Hegel puts it: "The rest of Nature is still not present for [the plant] as outer."[183] Animals (and particularly the "higher animals") are far more individualized; the gathering of their organs within a unifying skin expresses a degree of inwardness or subjectivity (a simple animal such as a sponge or echinoderm has no such unifying skin and thus remains plantlike). Yet we do not call even the higher animals *individuals* in the sense we mean it of self-determining human beings. The animal does not say "I." The individualizing principle (the animal *entelechy*), which is creative of animal form and behavior, reaches the level of soul or sentience, but not selfhood.

With the "higher" animals the developed, fiery soul-character extrudes and becomes directly perceptible in the manifestation we call the "voice" or "cry." The Goethean biologist Hermann Poppelbaum speaks of the "warm voice" of the mammals, produced by a stream of warm air. This he distinguishes from the "cold voice" of the less individualized lower animals — for example, the chirping of insects caused by friction of their limbs being rubbed on their abdomen.[184] We could also look at animal structure and behavior — the colors, patterns, the reproductive processes, even the manner in which animals die; in all these ways the soul expresses itself through the sensible form of the animal. It is through a Goethean form of qualitative perception that all such outward characteristics become "windows" to a particular quality of animal soul or animal inwardness.

In the higher plants we saw the three-fold principle expressing itself in the polarity of root and flower with the rhythmic leaf formations relating the two. With the animal this three-fold principle has turned inward, and through the animal kingdom it is progressively manifested in the animal's physiological organization. In the mammals this organization has become clearly expressed: one pole is the "sense-nerve organization" focused in the brain, the other is the "metabolic-limb organization" focused in the organs of metabolism and movement. Between these two is the "rhythmic organization" focused in the respiratory and circulatory systems. The polar nature of the sense-nerve and metabolic-limb organizations can be recognized in many ways. For example, the brain is relatively "cool" and motionless (held aloft in the skull) and its constitutive tissue non-regenerative; that is, relatively dead and close to mineral nature, thus expressing Earth. This is opposed by the warmth, immense vitality, generative activity and physical movement of metabolic and limb processes, expressing Sun. The rhythmic activities of blood and breath move between and unite these polar organizations and have characteristics of both. Their rhythm, for instance, is an oscillation between movement and momentary stasis; one phase of their rhythm carries "life" (oxygen), the other "death" (carbon dioxide).

Each of the three physiological organizations — the "sense-nerve," the "metabolic-limb" and the "rhythmic" — is related to a specific soul quality. In any individual mammal — say, a monkey — this threefold organization is present, but if we take the different mammalian groups and compare them, we find that in each group one or the other organization predominates. For example, the mouse belongs to the rodent group, which emphasizes the sense-nerve aspect. The rodent is alert, all outward, acutely awake to its environment; it is immensely sensitive and highly mobile or light-footed. In the soul quality of its shrill cry we hear an outward expression of its sense-nerve character. By contrast, members of the ruminant group such as cows are ponderous and slow, engaged intimately with the Earth in their processes of rumination and digestion. The ruminant is relatively unaware of the outside environment; its soul forces are drawn to an inner awareness of the process whereby vegetable matter is transformed into milk. The deep, slow mooing of the cow is an outward expression of its predominantly metabolic character. Between rodents and ruminants are found the carnivores — the lions, for example; these mammals have a soul character that oscillates between the nervous alertness and readiness of the sense-nerve type and the slow digestive sluggishness of the metabolic-limb type.[185]

Actually, we need to be able to inwardly picture an even more complex metamorphosis of gravity-levity (Earth-Sun) if we are to rightly understand the evolution of the animal (and even more so the human being). The inward substance and form aspect of the sense-nerve system of the animal indeed expresses Earth; but, in terms of function, it is the very "dead" mineral quality of nerve substance which allows sentience to shine forth and awakens the animal outwardly to its environment. So we must picture an "inversion" of Sun into Earth and the Sun principle thereby "finding itself" as the "lightness" or levity nature of sentient life (and the illuminating power of conscious thought in the human being). The inverse applies to the polar aspect of the animal, the metabolic-limb system. The inward substance and form aspect of the metabolic-limb system expresses the warmth and vitality of Sun. Yet the outward function of this system is Earth orientated, something we see clearly in the linear geometry of the limbs of the higher animals. This Earth orientation is also revealed in what these animals do or how they behave in relation to their environment (and in the power of will in the human being).

The animal can now be represented in a similar way to the geo-solar and plant organisms, as the expression of a threefold principle. We must, however, bear in mind the complex way in which Sun and Earth unite and transform each other in each dimension of the animal being. Soul, or sentience, cannot be graphically rendered — and neither can animal behavior. Nevertheless, for the purpose of indicating the metamorphosis of the "archetypal landscape" through the kingdoms of nature, a manner of graphical rendering is presented here. The three organizations — geo-solar, plant, and animal — represent a graded sequence that can be worked through with a fluid, imaginative thinking.

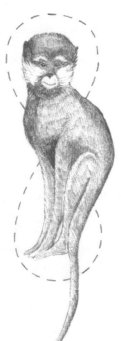

Earth and Sun inverted

Rhythmic organization

Earth and Sun inverted

and transformed into sense-nerve organization

and transformed into metabolic-limb organization

Fig. 22. The animal as the image of the archetypal landscape (gravity-levity)

Each organization is a moment or *gestalt* of a single evolutionary movement, just as the different leaf forms up the axis of the plant are moments of the unitary formative movement of the plant *entelechy*. What Goethe meant by heightening or *Steigerung* in terms of plant metamorphosis applies equally to the developmental process we call evolution. Just as the living idea, which is more or less hidden in the form of the leaves, is revealed in the flower, so the idea, which is more or less hidden in the geo-solar organization, is progressively revealed or realized in plant and animal organizations.

In this way we come to a method for understanding evolution, a method which is precisely the reverse of the procedure evolutionary biology conventionally employs. As we have already discussed, conventional biology, seeking explanations for the form and existence of all living earthly entities, follows their ancestry back into the past, where their origins are sought in their first or most primitive ancestor, the "progenitor organism," as Darwin called it. The mechanistic logic of causation requires it to work in this way — the cause must precede the effect. A creative approach to evolution strives to understand the laws of organic formation through an inspirative form of thinking which grasps that, in formative processes, time also streams from the future. This means that the form of the more "primitive" can only be understood in terms of what comes later, in the light of what is more "evolved." In relation to Goethe's notion of *Steigerung* we could say: only by grasping the "realized idea" in nature, as it manifests in the so-called "advanced" or "evolved" forms, are we able to rightly see and understand the "primitive" forms in which the living idea is more or less hidden.

Goethean biologists have studied specific aspects of evolutionary development according to such imaginative and inspirative insight. For example, Gerbert Grohmann explains that the gymnosperms and cycadophytes can in no sense be regarded as ancestors of the angiosperms but rather as the forerunners, the "prophetic forms" of the angiosperms. He shows that the forms of the gymnosperms and other primitive groups can only be understood *in terms of* the angiosperms.[186] Hermann Poppelbaum has taken the same approach in his exploration of the evolutionary relationship between the human being and the mammals, and Ernst-Michael Kranich has thoroughly explored this idea in relation to the fish and other vertebrate groups. Most recently, Jos Verhulst has demonstrated that, in his words, "something like an Aristotelian final cause must be at work in human evolution."[187]

The Human Being and the Evolution of Landscape

The modern scientific view of evolution is that the forms of life on this planet are in flux or in a constant process of adapting to changing environmental conditions. Evolutionary science opposes the religious conception, which sees life-forms as created "in the beginning" — complete and final. Physiologically, humans belong to the realm of biological life; our bodily organization has been created by the same processes that have created the physical forms of plants and animals. But human beings also belong to the realm of mind or spirit; we can grasp in thinking the laws or principles that are creative of organic form. In this sense we relate to the evolutionary process as co-creators. The revolution in genetic engineering, for example, is bringing home the fact that human beings now have the capacity to create new life-forms. Evolution, in a very definite sense, is moving into human hands.

Our investigation, through imaginative cognition, of the geological, plant and animal organs of landscape, revealed that a three-fold principle of living organization has progressively expressed itself or come to presence. As we have seen, the three different kinds of mammals — the rodents, the carnivores and the ruminants — express a predominance of sense-nerve, rhythmic and metabolic-limb qualities respectively. Every mammal (like every other animal) is highly specialized in its physiological organization and behavior with respect to the particular environment or landscape in which it evolved. What is true of the three mammalian groups viewed as a whole is true of every individual human being. In the human being these three systems are balanced, none becoming one-sided in expression. We "see" the complete human being if we consider the mammals as a whole. In other words, the human being makes concrete and immediate the principle of three-foldness, which is still more or less "dispersed" in the mammals (and far more "hidden" or less realized in the lower animal types).[188]

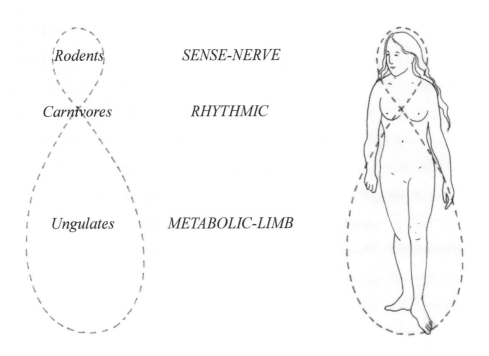

Fig. 23. The human form as the metamorphosis of the mammalian groups

It would be patently nonsensical to suggest that the flower is superior to the seed or leaf. Likewise it is possible to say that the human being stands at the summit of terrestrial evolution without any suggestion that we are superior to the other forms of life. What is meant, is that the human form is the fullest or most realized, individualized and balanced expression of the three-fold life principle which is one-sidedly expressed (or specialized) in the animal forms. This is the reason why Goethe (following Herder and others) declared that the animal kingdom is the spread-out human being and that, conversely, the human is the whole of the animal realm gathered up or individualized. We understand the human form through Goethe's notion of *Steigerung*: the human form is the "immediate truth" of animality. This means, as indicated above, that the animal realm can only be rightly understood *in terms of* the human form; the lawfulness of the creative or formative process in evolution relates to the stream of time that works from the future. This has another implication which stands in stark contrast to neo-Darwinian thinking (but does not necessitate a return to old religious conceptions): the whole thrust of evolution can in no sense be regarded as random. Only chance events, taking place in the stream of time from past to present (as all causal processes do), can possibly have the character of arbitrariness.

But we must go further than physiological form to find the human essence and the meaning of the human being in terms of the whole evolutionary process. The highest flowering of evolutionary movement takes place in human thinking. In thinking the organic is intensified into the spiritual or what Goethe called the "spiritual-organic." Through human thinking (and in the language which is the expression of that thinking), the manifold ideas in nature (hidden in the physical forms of things) achieve their spiritual form. The ideas in nature are concrete and substantial, although, as we have seen, they are "spiritualized" progressively in the flower and animal soul. When the human being understands a living form, that form undergoes a dramatic metamorphosis, an immense evolutionary leap. Through human thinking the forms of nature take on an entirely new kind of existence — they become *meanings*. Thus Goethe (with Aristotle) considered the state of "being known" to be a further stage of the phenomenon itself.[189]

We are now in a position to understand the relationship of the human being to the landscape in a creative way. This is how the Goethean ecologist Jochen Bockemühl expresses it:

> In the human organism, the individual presents himself directly to the observer through the different forms of expression. Organs and internal functions remain hidden from view. In a landscape it is the other way round, as the organs and their

functions are immediately perceptible, whereas the expression, language and unity of the whole are only found by bringing things together in inner vision.[190]

The language of the landscape is a gestural language. The landscape is obviously not *itself* speaking this language; rather, the language is concretely (or objectively) expressed in the organs of the landscape, which are spread out or external to each other in the visible forms of rock, plant and animal. Only through the thinking or inner vision of the researcher — that is, through some kind of phenomenological research — does the landscape actually come to *speak*; thus we can say that human speaking (or other creative expression) is the most heightened form of the phenomena. In the landscape the organs are visible, the "language" hidden. It is the other way round with the human being; here the principal organs are contained within a skin, and what is most outwardly expressed is language. With the plant, seen as an intensified expression of a landscape, organ-nature has achieved a degree of individuality, and the plant correspondingly realizes a degree of "language" — the concrete language of the form of the flower. The human being is, in this sense, the inverted landscape, landscape turned inside out, the most intensified dimension of landscape through which the landscape as a whole can express itself. This is the deeper organic connection between the human being and landscape which only imaginative-inspirative thinking can light upon.

The fact that an idea *is* a phenomenon of nature in a more developed stage of itself is something that remains entirely concealed at the level of Earth thinking, where an idea is seen as merely a mental correspondence to a thing or event. Hidden, too, are the moral responsibilities which derive from realizing the relationship between organic form and human thinking. Conventional ecology has a dimension of environmental values and ethics, but these values and ethics are brought from beyond the actual process of the research, pressed upon it, as it were, from outside its methodology. The common argument is that science *should be* more responsible, that human culture *should be* sustainable. By contrast, to engage in the Elemental pathway toward a Fire understanding is to participate cognitively in the creative dimension of landscape to the point of a moral realization. At the Fire level it is accurate to speak of the researcher as becoming the "voice of the landscape" and the research work and its applications as being a continuation of the creative processes at work in that landscape.[191]

For "I" is not just the separate "me" who is surveying, experiencing and ever seeking to penetrate a world "other" to myself. Objective thinking, thinking in terms of the division of subject and object, is a necessary step in the realization of a greater personhood. The Goethean science of living form is a spiritual path to the degree that it calls the researcher beyond the experience of the isolated self, which is the basis of Earth cognition. This pathway leads the researcher through the personal "me" toward a greater realization of individuality and responsibility. The stage of Inspiration is a conscious relinquishing of the personal ego-sense for the sake of another, because what is required here is full participation in the life of another being. Inspiration means a turning and a relinquishing. With Intuition the "I" knows itself as nothing other than the living idea which breathes within nature as its life — surging and blossoming, creative of form, burning and lighting up as human truth-saying. Human thinking becomes the creative Fire-essence of the landscape, its *logos* or creative word, and in that way landscape realizes its spiritual form. The Goethean researcher reaches out cognitively into the immediate otherness of place, and when insight becomes distilled and heightened it expresses itself in deeds of love, bringing something good and beautiful into the world.

CHAPTER 4

THE YABBY PONDS:

A GOETHEAN STUDY OF PLACE

Introduction

This Goethean landscape research was carried out in a bush environment adjacent to a body of water called Pittwater, which is an arm of Broken Bay and which lies some twenty kilometers to the north of Sydney, Australia. A large part of Pittwater is bounded by the Ku-ring-gai Chase National Park, and the study took place in a particular landscape within that larger environment — the terrain surrounding the "Yabby Ponds." This is the local name for a series of waterholes linked by a creek that is the final section of Salvation Creek, a minor water-course that has two main tributaries and that flows into Pittwater at Lovett Bay.[192]

The area of the Yabby Ponds is characterized by steep sandstone hillsides rising above the waters of Pittwater, leveling out to rolling bush country which abounds in flora and fauna. This location was chosen partly because of its quality of wilderness; the aim was to explore relationships between entities that are endemic to this place.[193] The sculptural rock formations in and around the Yabby Ponds and the plateau-like hilltop, with its dry sandy soil and hidden ponds and streams, make a powerful first impression. In this abundantly vegetated terrain, one is also struck by the highly varied shapes of tree trunks and foliage, the floral colors — especially in spring — and constant, multifarious bird calls.

The study documented here begins with a consideration of the general topography, the rocks and waters of the Yabby Ponds landscape, then proceeds to the plant and animal life. This approach has a definite aim which accords with the discussion of the previous chapter: it is to gain an understanding of how the archetypal landscape, or creative potentiality of landscape, expresses itself in a definite way in this landscape. With our phenomenological method we begin with Earth, with a study of the particular aspects or parts, and proceed toward Fire or an understanding of the wholeness which is creating the parts.

The study is by no means intended to be a complete ecological investigation of this landscape. The essential purpose is to demonstrate a way of working with the pathway of the four Elements. It is unequivocally the case that the results arrived at here do not represent final or absolute truths. At most, what is presented are glimpses of the unity or wholeness of this landscape. These results are preliminary, movements toward imaginative, inspirative and intuitive ways of understanding, which can be taken up and developed by others. It also needs to be stressed that the poems, drawings and paintings presented should not be taken as ends in themselves; the intention is only to indicate ways in which the arts can be brought into the service of a Goethean science of living form. These artworks do not prove anything about this environment in the sense of a conventional scientific explanation. This research follows a methodology which distils the physically perceptible and logical to the "plastic," then to the "musical" and the "poetical." What is being striven for through this methodology is an exact artistic embodiment of the unity dimension or wholeness of this landscape, not its explanation.[194]

First Impressions

This landscape has a starkly elemental quality — the wide waters of Pittwater, often darkened and agitated by rains and winds which break in from the nearby Pacific Ocean; the sculpted sandstone outcrops on the hillsides hardened and colored with iron; the vegetation periodically burnt out by bushfires leaving blackened branches and trunks long after new growth has sprouted; twisted trees and bushes. A tough landscape, wild and ancient.

One can approach the Yabby Ponds by walking along tracks through Ku-ring-gai Chase National Park, over broad and undulating hilltops covered with a low desiccated scrub, dipping down here and there into hollows and watercourses. Or the approach can be made across water, over the sometimes glittering, sometimes wild expanses of Pittwater. At its southern end this large inlet folds into bays, each with the hills of the National Park rising up close behind. Along the shores there are pockets of semi-rainforest in gulleys where creeks enter. Large mud-flats also appear in some of these bays at low tide, with clumps of mangroves poking their roots up through the mud. Around the shores are stands of eucalypts and casuarinas which rise tall above the understorey of ferns and bushes on steeply sloping ground. Higher up, passing through the sandstone outcrops around the rim of the hills, we enter an entirely different domain, the world of the Yabby Ponds.

This is not a mountainous terrain; here there are no soaring peaks that thrust up and fill one with a sense of grandeur or the sublime. Rather, there is the impression of being in an ancient, worn-down landscape. On the hilltops one feels exposed to a broad expanse of sky and at the same time connected to the spreading expanse of the landscape, with its folds and ridges and its secret water courses. On these hilltops, around the Yabby Ponds, a sense of stillness and mystery reigns, a mysterious elemental presence.

The writer D. H. Lawrence had a first impression of the Australian landscape when he visited the country for about three months in 1922 — and it terrified him. Lawrence described his impressions through the mouthpiece of the characters in his novel *Kangaroo*. He found it nothing like the gentle and accessible landscapes of Europe; "... so hoary and lost, so unapproachable" is how he described it. For him the landscape was ancient and somehow non-human; he felt the "roused spirit of the bush" like "a terrible ageless watchfulness."[195] This first impression never left him even when he was able to understand and love the bush; he was never able to overcome a sense of unbridgeable distance between himself and the Australian landscape. "You feel you can't see — as if your eyes hadn't the vision in them to correspond with the outside landscape," he wrote.[196] He experienced it as a landscape which is "too deep down," which does not want to be known, which is "waiting." Lawrence felt that his inability to penetrate to the secret of the Australian bush could not be put down to his unfamiliarity: "Nobody could get at it," he declared.[197]

Something of this mysterious, ageless presence is experienced around the Yabby Ponds. The twisted trees and dense scrub, dry and fire-blackened, have a hidden richness of life. Cool air floats over the scrub from the hollows where the pools are secreted, and the connecting stream trickles along its course before it reaches the waterfall and the abrupt descent to Pittwater. In this landscape of tough, prickly vegetation there is an abundance of small flowers of multifarious colors and forms. One of these is *Hibbertia linearis* — softness and warmth, a golden strength exuded in the midst of the desiccated starkness of the bush. A gentle quality, like a melting of the harshness. The hushed air is filled with the twittering of many small birds that are continually appearing and disappearing within this landscape. One of these occasionally flits out, issuing a fluid bell-tone which momentarily focuses the stillness — a darting, shadowy presence within the dense vegetation. This is the white-eared honeyeater. In the purity of the bird calls, in the tiny blossoms and sheltering warmth of the vegetation, one feels an intimate quality, a quietude, that contrasts with the vastness of Pittwater and its semi-tropical shores below.

Earth Cognition

Earth cognition — the geological organism

Climate

The landscape in question is located at approximately 33 degrees of latitude south of the Equator and has a temperate climate. The summers here are consistently hot and mildly humid, and temperatures can rise as high as 40 degrees centigrade, tempered by sea breezes (the sea is only a few kilometers distant). Winters are mild with temperatures rarely going below 10 degrees centigrade. During the spring in particular there is often heavy rainfall, with tropical-like deluges that can sometimes last for days on end. Very hot days in summer often end with strong cooling winds called "southerly busters." Otherwise the area is not characterized by winds of any particular kind.

Topography

The overall form of the terrain is undulating hilltops and ridges dropping to large bodies of water which are extensions of Broken Bay. The hilltops tend to be broad and relatively flat and to fall quite sharply to creeks and bays below (i.e., Pittwater in the immediate vicinity of the Yabby Ponds). On the ridgetops there are numerous large, flat sandstone outcrops, and on the hillsides there are small cliffs and irregular sandstone formations, commonly bulbous or rounded in shape. Small caves and overhangs have been hollowed out by the wind.

The sandstone terrain of the Yabby Ponds is part of the broader Sydney Basin formation. The main inlets in this basin — Pittwater, Broken Bay, Botany Bay and Sydney Harbor — are geographically regarded as "drowned river valleys" or *rias*, characteristic of the coastal terrain in this part of the world. These particular *rias* were formed around 6,000 years ago, through the Holocene rise in sea level, the most recent rise in the Earth's geological history.[198]

The Yabby Ponds themselves are a series of small pools joined by a winding streambed on a broad ridgetop above Pittwater. The pools are all located in a broad depression of the hilltop in which the stream moves slowly before falling about 30m and then flowing into Pittwater. The movement of water in these pools is seasonal; they can sometimes almost dry up during hot months.

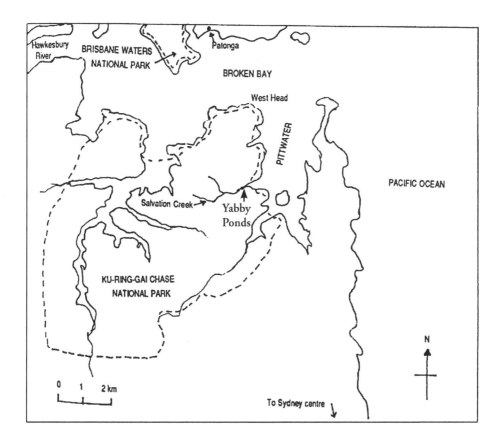

Fig. 24. Map of Sydney Basin/Broken Bay indicating location of the Yabby Ponds

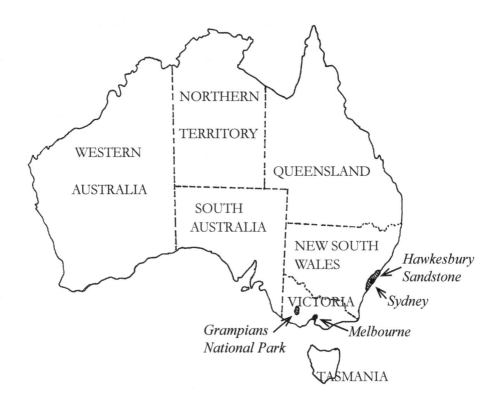

Fig. 25. Map of Australia showing location of the Hawkesbury Sandstone region (NSW) and the Grampians National Park (Victoria)

Rock and soil types

There is one rock predominant in the Yabby Ponds environment and this is a red/purple-colored sandstone called "Hawkesbury Sandstone." This rock-type actually occurs in a wide area which takes in the entire Sydney Basin, most of Broken Bay, the Hawkesbury River and its tributaries. Less common (but found in the Yabby Ponds environs) are red claystones. (Most of Australia's surface is covered with Paleozoic and Mesozoic sandstones, the most famous red-sandstone formations in the central region of the country being the Olgas and Uluru (Ayer's Rock).

Hawkesbury Sandstone (see Color Plate I, p. 133) is composed of sand (silica dioxide) grains in a softer matrix of iron oxide and iron carbonate. It was originally formed from siliceous detrital sediments — sediments from weathered igneous formations that were transported in water as solid particles (i.e., not dissolved) leading to sedimentation (alluvium). Deposited particles were converted to rock by cementation and compaction with iron compounds that precipitated out of water percolating through the deposits.[199]

The general characteristics of the sandstone are as follows:

- Hardness — 7 (Moh's Hardness Scale)

- Color — sand particles opaque to translucent white, matrix is darkish purple/red.

- Streak (color when powdered) — purple/red.

- Lustre — dull.

- Clastic texture — a coarse but not jagged texture; even-sized and rounded particles, the size of moderately coarse sand grains.

The soil type in the immediate vicinity of the Yabby Ponds is sandy — pale reddish-cream, sandy and very porous. It is very dry with little humus. In the places where the clay stone appears, the soil is more reddish and more clay-like.

Geological history of the Hawkesbury area

The history of the rock formations of this area has been subject to a number of interpretations.[200] There are three distinct rock strata which are typical of the area called the Sydney Basin: firstly, there is a series of sandstones, clay-stones, shales and conglomerates which is known as the Narrabeen Group. Above this is a smaller stratification of sandstone known as Hawkesbury Sandstone, upon which the city of Sydney is built; above this is the Wianamatta Group of shales. All three groups are the result of sedimentation processes, initiated and controlled by tectonic forces.[201] Earth movements probably gave rise to a plateau formation in the Permian period, forming a drainage system of streams which in turn allowed for the transportation and deposition of sand and pebbles over the breadth of the Sydney Basin area. Since Triassic times the rocks of all three groups have remained above sea level, and, as a consequence, they have been through very long periods of erosion.[202]

The Narrabeen Group was laid down in the Late Permian to Middle Triassic periods. There was presumably a great lake in the Sydney Basin area in the Late Permian period, connected with a Triassic lake system in the inland of Australia. As the area subsided, this system alternated between swamp and lake into which sediments were deposited, resulting in the

Narrabeen Group of sedimentary rocks. This group is up to 760 m thick below Sydney. The shales of this group, formed from mud deposited in this lake/swamp system, contain many particles of copper and volcanic ash (showing that volcanic activity was taking place in the vicinity). In the middle depositional period of the Narrabeen Group, in the north Sydney Basin area, clay-stones were formed and colored into red shades by iron that was oxidized in situ.

On top of the Narrabeen Group were deposited the sediments that became Hawkesbury Sandstone, the rock found at the Yabby Ponds site. The quartzose sands which make up the sandstone were probably derived from upper Devonian quartzites of a more distant origin as well as from granites that were intruded in the Carboniferous and Permian times and that would have become exposed. Granites contain a high percentage of silicon; quartzite is a rock of a wholly siliceous nature.

Earth cognition — the plant organism

Overall characteristics of the vegetation

The vegetation of the Yabby Ponds environs (representing the Sydney region flora in general) shows an enormous variety of plant species, representing many different families. The area is one of the floristically richest in Australia, with about 2,000 indigenous species.[203] In this sense it can be compared to the Grampians (sandstone) Ranges in Victoria which contain about one third of that state's plant species (see Figure 25, p. 100).[204] Ten-meter transects in the Yabby Ponds area yield around 20 species of shrubby plants and many herbs.

The vegetation is characteristically xeromorphic, reflecting the dry, porous nature of the sandy soils and the sandstone. In the immediate vicinity of the Yabby Ponds, the vegetation is dense heath mixed with stands of stunted eucalypts, banksias and larger angophoras. On the hillsides eucalypt vegetation forms a dry sclerophyll forest type.

A transect taken at the Yabby Ponds represents only a sample of the species present (see Figure 26). It shows a grading from water plants to those which grow in the dry expanses of heath with taller but stunted eucalypts. *Hibbertia linearis* (one focus of this study), sometimes known as the showy guinea flower, shares the understorey with shrubs such as *Telopea speciosissima* (waratah), *Grevillea sericea* (silky spider flower) and *Grevillea buxifolia* (hairy spider flower), *Actinotus minor* (lesser flannel flower), *Dillwynia retorta* (bacon and eggs), *Boronia ledifolia* (*Ledum boronia*) and *Acacia ulicifolia* (prickly moses). The tree species in this location are predominantly *Banksia serrata* and *Eucalyptus piperita*.

The families

The Australian landscape in general is dominated by the genus *Eucalyptus* which is a member of the plant family Myrtaceae. In the Yabby Ponds environs *Eucalyptus* is not so strongly expressed, but rather many heath-like genera and a number of water plants. The heath plants are commonly members of the family Proteaceae (i.e., *Banksia, Persoonia, Hakea, Telopea*) but there are numerous other families represented as well, including Epacridaceae and Fabiaceae.

Hibbertia linearis

Hibbertia linearis individuals are profuse in the Yabby Ponds environment, springing up in the wake of bushfires which swept through the area during the decade preceding this study. Growing in patches, they sometimes dominate the understory — many are relatively immature plants, whose numbers will quite likely be diminished when the bush becomes fully established again. *H. linearis* plants flower profusely in the spring (August-October), but some flower throughout the year, though less in the summer to autumn months (January-March).

The roots grow vertically downwards and have a dark brown skin over a paler woody interior. About 4 mm in diameter near the surface of the ground, they taper down gradually, sending out lateral rootlets around 200 mm apart, each about 1 mm in diameter and of a pale color.

The lower main stem (axis) of the immature plant is of a pale reddish/yellow color, covered with darker red bark-like fibers vertically orientated, somewhat woody but flexible. The main axis is around 5 mm in diameter at its base and rises vertically, gradually tapering to the highest point where it is around 2 mm in diameter. The central stem of the individuals studied rises to about 1.5 m. The highest parts of the stem and the young shoots have no bark-like fibers, are of a crimson color and are covered with a fine short white hair (tomentum). Around 30 cm from the base smaller stems branch off from the main stem in spiraling fashion, in alternating positions, the longest on the juvenile plants being about 50 cm.

The first leaves appear around 20 cm from the base of the plant. The leaves are attached to the main stem by woody petioles about 1 cm long, the leaves being more or less bunched on the first lateral stems, but, on the older ones, alternating at spaces of about 1 cm and spiraling. At the termination of most lateral stems the leaves form a whorl. Each leaf has a dark crimson base and a central vein of darker green color. The longest leaf is around 3 cm, the shortest 1 cm; all are about 4 mm wide at the widest point, generally linear-oblong to obovate in shape. No metamorphosis of

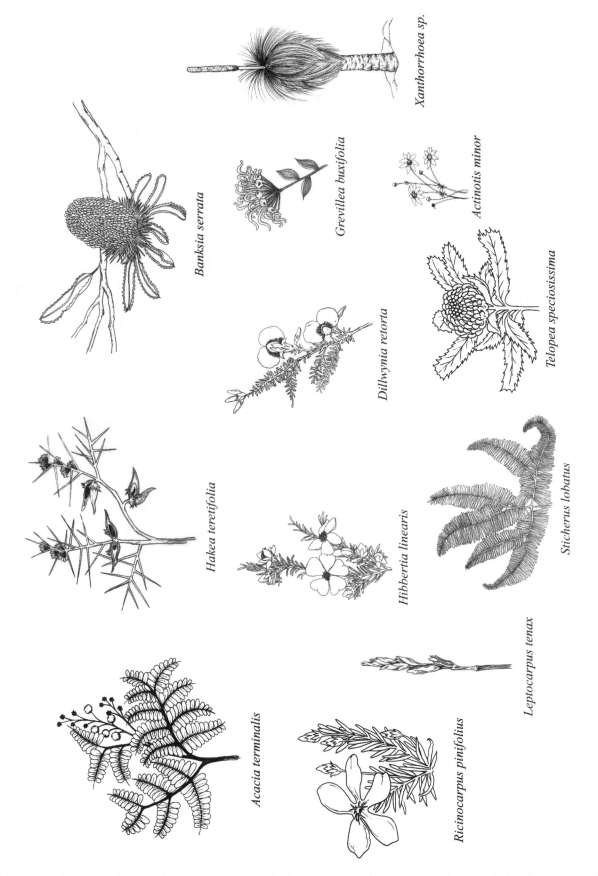

Xanthorrhoea sp.

Banksia serrata

Grevillea buxifolia

Actinotis minor

Dillwynia retorta

Telopea speciosissima

Hakea tereifolia

Hibbertia linearis

Sticherus lobatus

Leptocarpus tenax

Acacia terminalis

Ricinocarpus pinifolius

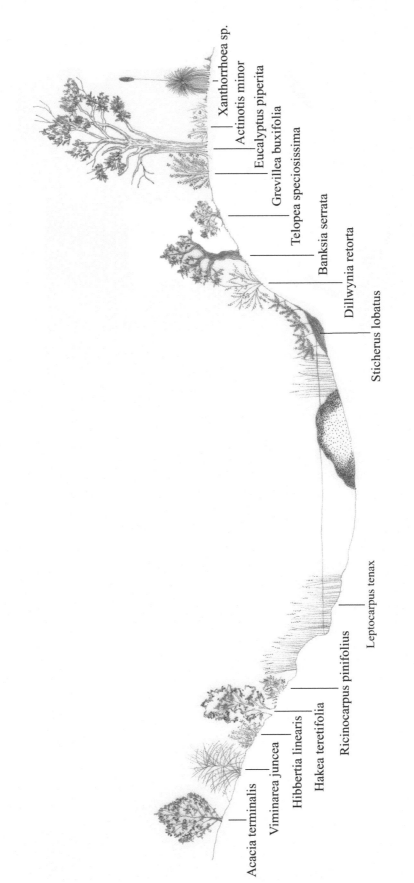

Fig. 26. The Yabby Ponds: typical plant species. Above (p. 104): individual species; below (p. 105): a cross section

leaf shapes is discernible from the base to the top of the plant. The upper surface of the leaves is olive green, with no hair (glabrous) and a smooth waxy texture; the under surface is pale green to white, roughly textured, with the leaf margin curved slightly downwards (recurved). The leaves are virtually tasteless.

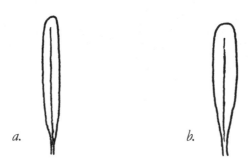

Fig. 27. Leaf shapes of Hibbertia linearis: *a) linear-oblong;*
b) linear-obovate

Flowers arise both terminally on the lateral stems — at the center of whorls of leaves — and in leaf axils on the lower main axis. There is no flower stalk (they are sessile). At the base of the inflorescence there are five pointed, light green sepals, at the center of and superior to which there are two onion-shaped carpels, about 2 mm long and attached to each other at the base, with a green stalk (style) emerging from the center of each. Each carpel contains two white globules (ovules), each less that 1 mm in length. Growing from the bases of these carpels, and surrounding them, are about fifteen bright golden stalks (filaments), around 5 mm long, with a 2 mm-long thin golden head (anther) at the end of each. Also arising from the bases of the carpels are five golden, heart-shaped petals, each about 1.7 cm long, and arranged in a very open cup. The flowers have no conspicuous scent or taste.

Fig. 28. Hibbertia linearis *— cross-section of flower showing two carpels, each containing two ovules*

The fruits are dry, woody, bi-partite structures, each part being about 1.4 cm long and joined at the base to the other. The parts break open (dehisce) at the top, and each releases two seeds. The seeds are about 1 cm long, globular to kidney-shaped (reniform) and of a strong red-brown color.

The species *H. linearis* belongs to the family Dilleniaceae (the Dillenia family), genus *Hibbertia*. This family comprises ten genera worldwide, and 350 species, mostly within tropical America and Australasia but also Africa.[205] All are trees or shrubs, and genera include *Davilla* (35 species) and *Curatella* (2 species), from South America. *Hibbertia* (about 110 species) is the largest genus, and occurs in tropical to sub-tropical S.E. Asia and Australia.[206]

The Dilleniaceae family is the largest of two families which belong to the order Dilleniales, and the subclass Dilleniidae, class Magnoliopsida (dicotyledons). Subclass Dilleniidae is related to but more advanced than another subclass — Magnoliidae — the latter including the Magnoliaceae family which is considered to be one of the oldest families of flowering plants.[207] Dilleniaceae is considered to be relatively primitive in a number of features which relate it to Magnoliidae; these include the distinct or separate (conduplicate) carpels and the presence of numerous stamens.[208] In the key to the Angiospermae the Dilleniaceae family is identified as having free carpels and two or more ovaries; petals free or absent; styles free from one another; leaves not aromatic; stamens hypogynous (attached to the receptacle below the gynoecium); as being either trees, shrubs, or climbers with simple leaves; as having a yellow corolla which is conspicuous.[209]

The Dillenia family is of little domestic importance, but some species of *Hibbertia* and other genera are cultivated as ornamental plants.[210] In the Sydney region there are 24 species of *Hibbertia*, all shrubs or climbers, having conspicuous yellow flowers which are solitary and terminal or apparently axillary.[211] *H. linearis* is widespread in the Sydney region, occurring in heath country and dry sclerophyll forest, on sandstone and old dunes. It also occurs on the coast, tablelands and slopes of New South Wales and Queensland, and in Victoria and Tasmania.

Earth cognition — the animal organism

There are a large number of animal species in the region of the Ku-ring-gai Chase National Park in which the Yabby Ponds are located.[212] In the course of this study a number of these animals were sighted within a kilometer radius of the Yabby Ponds — from reptiles such as the lace monitor or goanna (*Varanus varius*), and the highly poisonous red-bellied black snake (*Pseudechis porphyriacus*), to marsupials (brush-tailed rock-wallaby [*Petrogale penicillata*]), to crustaceans (fresh water crayfish or yabby [*Parastacidae* sp.], after which these pools are named).

*Figure 29. a. Lace monitor or goanna (*Varanus virius*) (⅛ actual size);
b. Fresh water crayfish or yabby (*Parastacidae *sp.*) (½ actual size); c. Grey
fantail (*Rhipidura fuliginosa*); d. Spotted pardalote (*Pardalotus punctatus*)*

*Figure 30. a. Red-bellied black snake (*Pseudechis porphyriacus*)*
*b. Brush-tailed rock-wallaby (*Petrogale penicillata*)*
*c. Little wattlebird (*Anthochaera lunulata*)*
*d. White-eared honeyeater (*Lichenostomus leucotis*)*

The diversity of birdlife in the Yabby Ponds environment is great. At the bottom of the ridges and along the shoreline of Pittwater are found some large birds, including the sulphur-crested cockatoo, kookaburra and the rare black cockatoo (which feeds on the nuts of the casuarina trees), as well as water birds such as cormorants. The birds in the immediate heathland to low forest environment of the Yabby Ponds are smaller in stature and extremely varied in song type. These include the white-cheeked honeyeater (*Phylidonyris nigra*), little wattlebird (*Anthochaera lunulata*), grey fantail (*Rhipidura fuliginosa*), spotted pardalote (*Pardalotus punctatus*) and brown thornbill (*Ananthiza pusilla*).

The White-eared honeyeater (Lichenostomus leucotis)

This bird is the second focus of this study. It was common in the Yabby Ponds landscape at the main time of the study — i.e., spring to summer.

Lichenostomus leucotis (see Color Plate II, p. 133) has a body length of 205 mm, wing length of 102 mm, tail 96 mm, bill 15 mm, tarsus 25 mm (approximate sizes). It weighs about 22 g. The top of the head and the hindneck of the adult is dark grey with black streaks; most of the rest of the body, including the tail and wing feathers, is of a yellow/olive green color. The side of the face and down the breast is black with a white patch extending from behind the eyes backwards and down the neck. The lower part of the breast and underpart of the bird is a paler yellow/green color; the bill is black, slightly curved and sharply pointed; the eyes are dark brown to black, legs dark grey to black. The sexes are similar and there is no seasonal variation. The juvenile is similar to the adult but is of a duller green and the top of the head is green rather than grey.[213]

L. leucotis belongs to the family Meliphagidae or honeyeaters, whose main anatomical distinguishing feature is the brush-tipped tongue which is used like a paint-brush to collect nectar and other fluids by capillary action.[214] It extends well beyond the beak into tubular flowers and cracks in trees. Honeyeaters are grouped within the suborder Passeri according to the classification of Sibley & Ahlquist, based on DNA studies; they are "oscines" or "true songbirds."[215] Other distinguishing features of the honeyeaters are the usually dull green or brown plumage with brighter patches around the face, the curved bill and their nomadic habit.

There are 151 species of honeyeater in the world, 73 in Australia.[216] The honeyeaters are a highly successful bird family in Australia, often accounting for more than half the bird species in any one area. This can be attributed to the fact that they have a diversity of food sources (nectar from a broad range of plants and also insects) and their pugnacious habit — they actively defend their food supplies.[217]

Two races of white-eared honeyeater are recognized — a smaller dry-country form endemic to mallee habitats in Western Australia (*L.l. novaenorciae*), and a larger form (*L.l. leucotis*) endemic throughout eastern Australia. Throughout their range the honeyeaters occur in coastal heaths, woodland, wet and dry sclerophyll forest and mallee, with no apparent preference for a particular floristic association. *L.l. leucotis* appears to prefer a eucalypt overstory (no particular species) and a dense scrub understory for nesting.[218]

White-eared honeyeaters tend to be solitary, but defend territory in pairs throughout the year. They may form groups of up to six individuals. They have a characteristic habit of removing the hair from animals or wool from human clothes for their nests.[219] In the Yabby Ponds environment, the bird was observed to be constantly darting in and out of trees and bushes, occasionally resting on branches to preen or sharpen its bill or utter a cry, staying in one tree for only a short time before moving to another close by, occasionally flying to more distant trees in the environment. The birds were observed to be mostly solitary, but on occasions they moved in pairs.

This bird is not migratory but, in winter, moves to high altitudes from the coastal plains. In the Yabby Ponds environment, the birds were first sighted in the breeding season. Another study of the white-eared honeyeater in a location relatively close to that of the Yabby Ponds (at Brisbane Water National Park) found definite seasonal fluctuations of the populations of these birds, showing peaks of abundance in winter and low numbers in summer.[220]

Despite the name "honeyeater," the food-source of this bird is not only nectar from flowers. A part of its diet is insects and spiders, which it gleans from the bark on the branches and trunks of trees in the eucalypt canopy. On the occasions when the honeyeater gathers nectar and pollen from flowers, if the flower tube is too long for its beak, the bird simply pierces the flower.[221]

The territorial cry of this bird is distinctive and strong; it has been described as "a loud, double knocking 'chok-chok' or 'choo-choo' uttered infrequently"; also as a "loud 'chock up, chock up' and other softer calls."[222] Higgins identifies eight different types of call, the use of these call-types changing through the year. Several of these calls are infrequent, only two being heard frequently — those which he names the "chew" call (chew chew chew or choor choor choor . . .), and the "two-note calls" (tch-tchew, tch-tchew, tch-tchew . . .) — these two being used in the same circumstances.[223] Higgins suggests that the function of these two calls, used throughout the year, is to advertise ownership of territories, define territorial boundaries and maintain the breeding pair bond. The quality of these calls may be described as a strong, simple and liquid song (as opposed to, for example, an erratic chirp, a strident squawk, a chattering or a shrill twittering).

Higgins analyzes the call types by using a sonogram. The main territorial calls have a definite musical structure and one of these was worked out by the present researcher.

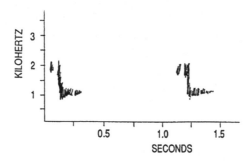

Fig. 31. Sonogram of a main territorial cry of the white-eared honeyeater (after Higgins, 1992)

Fig. 32. Musical form of a main territorial cry of the white-eared honeyeater

The bird breeds from August to December; both male and female feed and care for the young. Nests are constructed from July on, usually in the dense scrub understory and close (2-3 m) to the ground. The nest is made of a cup of grasses and twigs, lined with hairs or other soft material. Eggs are oval, 2-3 in number, white with red-brown spots at the larger end, incubating in fourteen days. The fledging period is also fourteen days.[224]

Water Cognition

Water cognition — the geological organism

In the previous chapter, Earth and Sun — taken together, in their dynamic relationship — were spoken of as the "body" of the geological organism, and the particular terrestrial environments as its organs. The Yabby Ponds landscape is a relatively tiny organ of Earth-Sun but, as with any organized body, even the smallest part gains its meaning only in relation to the greater whole. Water thinking is the first stage of getting to know this geological landscape in a participatory way.

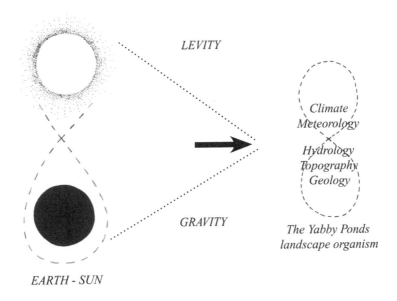

LEVITY

Climate
Meteorology

Hydrology
Topography
Geology

GRAVITY

The Yabby Ponds
landscape organism

EARTH - SUN

Fig. 33.

The method used is as follows: we enter into and "run through" the particularities of the Yabby Ponds environment, including the topographical character, the form of the sandstone, the shape and character of the watercourses, the climate and geological history of the area. This, in the first place, requires a strong familiarity with the Earth facts, which can be memorized or inwardly digested and then gathered together as a dynamic mental picture. There is no particular order in this process — overall we are not dealing with a developmental sequence. Indeed, one continually juxtaposes and re-juxtaposes different features or aspects in the process of building this mental picture. What follows is only one way of proceeding.

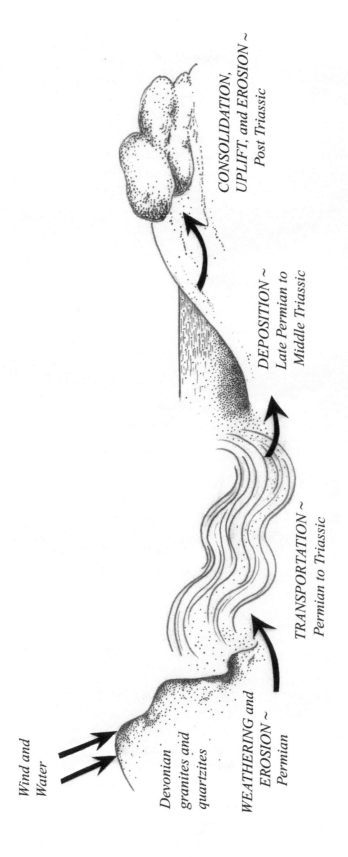

Fig. 34. Stages in the formation of Hawkesbury Sandstone

Moving imaginatively between the fine-textured claystones and the coarse siliceous sandstone (with its iron-red matrix and characteristic texture and hardness), we sense the relative substance-qualities of the rock types. Now, from the substance-quality of sandstone we move into the topographic features, forwards and backwards, building a fluid sensing of how the substance-quality of the sandstone expresses itself in shape (geomorphology), on scales large and small — the hollowed or bulbous rock formations and the deep hollowing processes forming the rias. Substance to shape, shape to substance — flowing from one to the other the active imagination weaves a dynamic inner picture of the geological organization.

Within this array of substances, forms and processes, there is a developmental sequence — the geological history of sandstone in this region. We now participate in the formation of the sandstone substance of which we have already built an Imagination; we inwardly experience each step and the translation of each step into the next (see Figure 34). Through the stages of weathering, transportation, sedimentation and so on, our already-developed Imagination of the physical "body" of sandstone now acquires a time "body" with a particular rhythmic character. In this way we are following the method by which Goethe himself contemplated geological formations — dynamically, always in relation to *how* rocks were deposited, *how* they arose.[225] Our dynamic Imagination of the formation of Hawkesbury Sandstone can now be expanded to embrace whole-Earth processes; we inwardly experience weathering, transportation and sedimentation as processes belonging to the geological organization of the Earth. We then "flow" back into the specific arising of Hawkesbury Sandstone in the Sydney region.

Now our dynamic inner picture of the geological character of this landscape — our living Imagination — is further developed through an imaginative experience of the characteristic climatic qualities and processes. Forwards and backwards, moving between substance, form and process, as well as through the seasons in this specific semi-tropical climatic zone, we participate in a comprehensive Imagination of the geological-geographical conditions of the Yabby Ponds landscape as an organ of the whole Earth organism.

Water cognition — the plant organism

As with the geological organism, the task now is to enter into and unite all the Earth facts concerning the vegetative nature of this landscape. An imaginative understanding of the vegetative landscape of the Yabby Ponds is built up by approaching this phenomenon from different sides, through participating in the different forms and through inwardly experiencing the different ordered sequences that have been established in the Earth stage. Again, what follows is only one way of proceeding.

We may firstly move imaginatively between the forms of the different plants which occupy the Yabby Ponds terrain, working with different juxtapositions, in pairs and larger groupings, sensing and inwardly picturing differences and similarities of form and of position in relation to the ponds and creek. From *Grevillea buxifolia* to *Hakea teretifolia* to *Hibbertia linearis* to *Actinotis minor* and so on, with Water thinking we flow through the whole population and in this way begin to develop an Imagination of the individuality of the population, of the plant organization as a dimension of the Yabby Ponds landscape. The different plants represent a number of different families, yet in their overall rough texture, their dry, reduced form and low, even stunted habit, they constitute a unified expression of the formative potential of this landscape. Diagrammatically this Imagination of plant population is pictured as a dynamic circle expressing the wholeness of the vegetative organization, with no particular sequence of plants intended.

The Imagination of the wholeness of this plant population can now be focused and intensified in the form of the particular species, *H. linearis*. First we imaginatively participate in the metamorphic sequence of the growth of this plant, by entering into and dwelling in the lower leaves and stem, then flowing into the other organs, upwards toward the flower, fruit and seed, moving backwards and forwards throughout the living body of the plant, taking in form but also texture, color, scent and taste (see Color Plate III, p. 134). This is the deep and lengthy process of getting to know this plant intimately. We experience this growth process of *H. linearis* from out of the whole landscape; firstly we picture the whole vegetation and then move to and through this species, then we move back again to the whole. In this way, *H. linearis* — with its low, open and relatively linear vegetative structure and small, strongly-colored flowers — is experienced and understood as an expression or focalization of the landscape as a whole.

This imaginative work with *H. linearis* can now be intensified in another way. To the extent that we have before us taxonomic information concerning the sequence — order Dilleniales, subclass Dilleniidae, class Magnoliopsida, family Dilleniaceae, genus *Hibbertia*, species *H. linearis* — we can imaginatively flow through this sequence, backwards and forwards, running through the general features which relate to the order, class and subclass (which belong to the whole-Earth organism but have no particular location) right down to the features that relate to the species in a particular landscape, inwardly experiencing the sequence as a fluid continuum. We inwardly picture the general characters of separate carpels and numerous

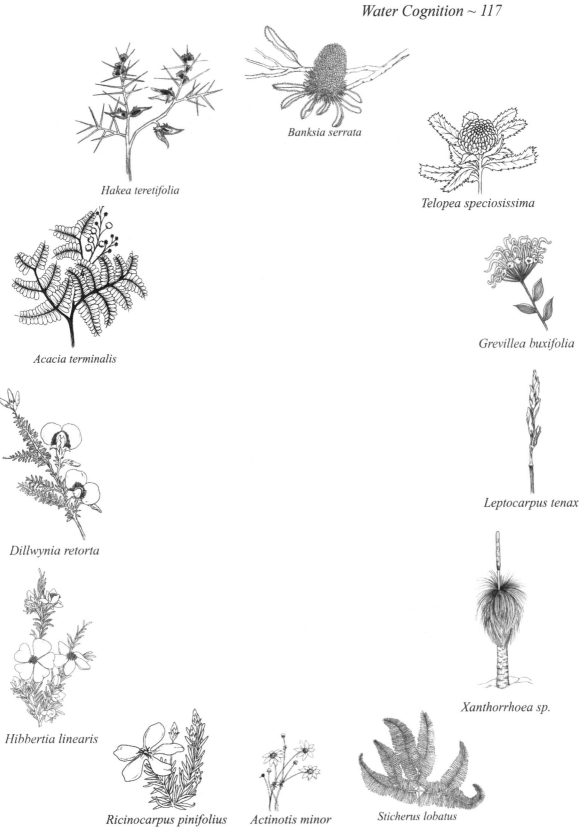

Hakea teretifolia

Banksia serrata

Telopea speciosissima

Acacia terminalis

Grevillea buxifolia

Dillwynia retorta

Leptocarpus tenax

Hibbertia linearis

Xanthorrhoea sp.

Ricinocarpus pinifolius

Actinotis minor

Sticherus lobatus

Fig. 35. A "circle" of plant population variation at the Yabby Ponds

stamens (which belong to the order and class), then move to the different shrub, tree and climber forms characteristic of the family and genus, thence into the particularities of the yellow-flowered shrub which is *H. linearis.* What normally are taken to be distinct, static taxonomic categories are now experienced as a living creative process.

The imaginative work at this Water stage does not, finally, picture the geological, plant and animal organizations as separate from one another. As set out on these pages they follow a particular sequence, but in practice we weave our Imaginations of the geological, plant and animal life of the Yabby Ponds into a comprehensive inner picture. For example, after participating in the formative processes of *H. linearis* through the family Dilleniaceae and genus *Hibbertia*, we "marry" this dynamic Imagination with our developed Imagination of the geological organization of the Yabby Ponds. We inwardly run between the two, sensing their relationship or continuity, experiencing the formative lawfulness whereby the particular plant — *H. linearis* — takes form in this geological landscape. In this way we are actually practicing Goethe's own way of working as discussed in Chapter 3, where it was noted: "For Goethe ... outer conditions merely bring it about that the inner formative forces come to manifestation in a particular way."[226] Slowly, by participating in this landscape in such ways, through entering into many different juxtapositions and sequences, we build up our imaginative picture of the Yabby Ponds landscape as a whole living entity.

Water cognition — the animal organism

An imaginative (Water) thinking is now brought to bear upon the animal life at the Yabby Ponds. As with the geological and plant organizations, using exact sensory imagination we "run through" the diverse facts and sequences presented in the Earth stage. The aim is not to test a hypothesis or establish a proof, but rather to allow the formative movements that have created this landscape to come to light. Ultimately, we seek an understanding of the relationships between the geological, vegetative and animal organizations of this landscape and insight into how the wholeness of place (the archetypal landscape) has "imparted" itself as its animal organization.

We start with an overall picturing of the animal life. For our purposes here it suffices to "run through" a sample of species found at the Yabby Ponds, just as we "ran through" the vegetation profile of plant species. The order of the animal forms given below is arbitrary in a spatial or temporal sense; this is not a metamorphic sequence but a sample of the variation of animal species in the landscape. Dwelling in the particulars of each animal type (to the extent that these are ascertained in this research), then running between the different animals in different orders, Water thinking begins to develop an overall fluid picturing or Imagination of the animal organization of this landscape. Through the form of the goanna, to the rock wallaby, to the red-bellied black snake, to the kookaburra and white-eared honeyeater and so on, the exact imagination begins to experience their belonging-together as aspects or parts of the same landscape-whole. These animals collectively are the landscape — or at least, one dimension of its manifestation. The goanna, rock wallaby and white-eared honeyeater are, of course, very different kinds of creatures; but, for example, in their relatively subdued coloration and color patterning, reclusive and (more or less) solitary habit and smallish stature we find a unifying tendency that relates them to the characteristic low, rough but enclosing or "intimate" quality of the vegetation. As with the plants, information is presented here in circular form to indicate that this is not a growth sequence but a manifestation of variation within the Yabby Ponds landscape.

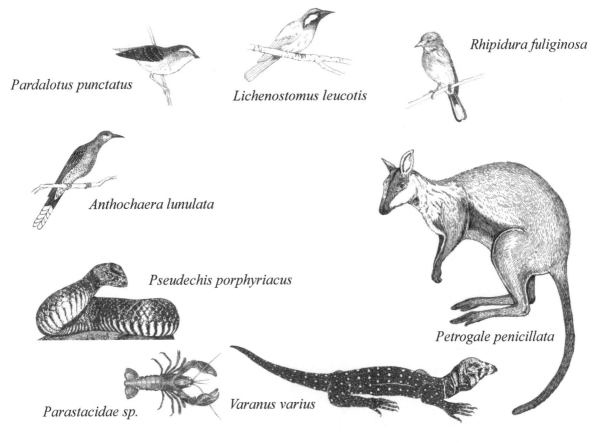

Pardalotus punctatus

Lichenostomus leucotis

Rhipidura fuliginosa

Anthochaera lunulata

Pseudechis porphyriacus

Petrogale penicillata

Parastacidae sp.

Varanus varius

Fig. 36. A "circle" of animal population variation at the Yabby Ponds

Now let us consider one significant example of the Yabby Pond fauna, the white-eared honeyeater. We will attempt to permeate in its entirety the factual "Earth information" concerning the species *L. leucotis* with the fluid activity of an exact sensory imagination. Here we do not, as with a flowering plant, have before us the ordered sequence of a growth process (i.e., from root, to leaf, to flower, to fruit and seed). Each piece of information concerning *L. leucotis* reflects a different aspect of the living being of that organism, but not in any sequential sense. Thus again, the information is presented in a circular form to show that we are working within the sphere of the whole organism. Water thinking dwells in each aspect, flowing then to the next in the organic "circle" — in practice, in many different orders or juxtapositions — until the totality of this information is gathered in a fluid inner picturing or Imagination. Thus we participate more and more deeply in the form and life processes of this bird.

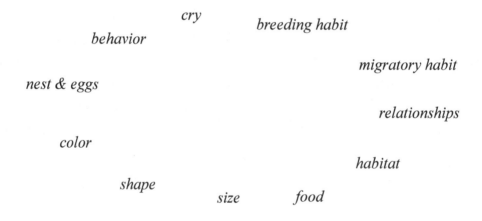

Fig. 37. White-eared honeyeater — the "sphere" of the whole organism

We move now with an exact sensory imagination through the order, to the suborder and family, to the species (*L. leucotis*), then back along this sequence towards the order. Just as with the plant groupings, imaginative thinking unites these bird groupings as a fluid continuum; we enter into the particulars of one ranking, then flow to the next. In this way Water thinking experiences the relationship of the general (which belongs to the whole-Earth organism) to the specific organism as a formative movement. We "live into" the character of perching (which belongs to the passerine order of birds in general), thence into the character of "true song," which belongs to the suborder, through the "brush-tipped tongue" character (of the family) into the features of the species which include its particular manner of perching and flitting and the unique form of its flute-like song. Thus Water thinking experiences a necessity in the relationship between perching and singing in the formative movement that has created the white-eared honeyeater.

Let us suppose that we have been able to develop a dynamic inner picturing of the formation of *L. leucotis* through the order and family, a living Imagination of this bird. With this in mind, we look back to our imaginative work with the geological and plant organizations of the Yabby Ponds landscape where we experienced, in the form of the rocks and different plants, its tough, desiccated, sculptural quality. Now, as we bring these different Imaginations together and move between them, it becomes possible to inwardly "see" a necessity in the arising of *L. leucotis* in this landscape as a living, creative process. At this point we are on the verge of an inspirative mode of perception.

Air Cognition

The wholeness of place is the organic lawfulness which reveals itself at first musically, as "toneless tone." In Air thinking (Inspiration) all forms and metamorphic sequences, all growth processes, become transparent to their formative musical gesture. Studied as a whole to the point of Inspiration, the living beings that comprise a place are realized as "movements" of a "symphony" of interrelated gestures, the wholeness of place being the comprehensive musical idea.

The diversity of Earth facts has been gathered up in the Water stage; Imagination perceives the formative movements that unite these otherwise separate details, ultimately the generative movement which has imparted itself as the geological, plant and animal organizations. The allowing-acquiescing which is Air thinking means allowing these Imaginations to be distilled, to become transparent to their musical idea.

Air cognition — the geological organism

The polar nature of Earth-Sun reveals the gestural character of gravity-levity — this has been discussed in the previous chapter. At the Earth pole the force of gravity consolidates, darkens and rigidifies, and at the Sun pole the force of levity expands and lightens; this gravity-levity polarity is united through a rhythmical interplay. The resulting three-foldness of levity-gravity united in rhythm is the essential "music" of the geological organization.

The rhythmic dynamic of gravity-levity which governs rock in its fluid condition (either molten or carried in a fluid medium such as wind or water) becomes "frozen" or solidified in rock formations. A dynamic morphology of rocks recognizes that every rock type is an expression of either gravity or levity or a more or less complex combination of the two; a rock, in other words, is a formative musical gesture come to rest. The task at the Air stage of Goethean phenomenology is to "read" the musical language of the rock types.

Fluid superheated rock (magma) originates deep in the Earth's mantle, from where it may rise and penetrate the crust. So, to the extent that a warming, lifting force is working on this primal rock substance, we can say that it is in a levity condition. When it begins to cool, the process of crystallization leads to solidification. Crystals form around innumerable "seed" centers through the process of accretion — a gravity process, predominating in the formation of igneous rocks. Crystalline, granitic rock constitutes a large part of the Earth's crust, including many of its mountain ranges, which were formed when this rock was uplifted by the cycling levity action of the semi-plastic mantle.

Goethe's poetic contemplation of granite led him to think of it as the primordial rock, the foundation of the Earth and essence of all solidity. He speaks of it as "the deepest and highest." Hegel calls granite "the heart of the mountain," the concrete principle par excellence. Metals, he says, "are less concrete than granite." It is granite's hardness, solidity, its depth and magnitude in the structure of the Earth, that makes a poetic observer like D.H. Lawrence sense it as a kind of primordial life which makes his "feet shiver."[227] The consistent character of such observations justifies us in speaking of a definite "granite image," which is the essential gestural character of this rock.

When solid granitic rock within the Earth's mantle is upthrust by levity forces on a large scale (orogenesis or mountain building), the exposed rock may be taken hold of by the levity powers of warmth, wind and water (i.e., weathering). Layers of rock peel off and the result is the regular rounded shapes (tors) which characterize granitic landscapes. Weathering is the agent of intensification or heightening (*Steigerung*) as it works in the geological organization. Through it, the idea or principle hidden in the unweathered form of the rock comes to expression. In the case of granite this "hidden principle" is the spherical form.[228] Granite, the primordial rock, the foundation of the Earth, reveals the spherical form of the Earth itself, a universal form — unindividualized, non-specific, the basis or potential for all other forms. The regular rounded form of tors shows a predominantly generative quality, meaning something near the beginning of a process of growth or expansion — seed-like in form, rounded, as are all youthful things in nature.

Fig. 38. Weathered granite (tors) expressing a "youthful," generative quality

Sandstone, the rock found at the Yabby Ponds, has moved further along the path toward the organic or specialized. It could be said that sandstone is granite in a higher levity state. Granitic rocks are worked upon intensively by the levity forces of warmth, wind and water; then the products of weathering (sands) are further "lifted" by wind and water, transported and deposited to form layers of sediment (gravity). The particles of the sediments are later cemented by means of uprising waters carrying iron compounds in solution which permeate and envelop the sand particles before fixing the particles within a matrix (a gravity process).

When consolidated sandstone is itself uplifed or otherwise exposed to the levity forces of wind and water, a hollowing process takes place which results in the gnarled, irregular sculptural formations that characterize sandstone landscapes. What comes to expression now is the most intensified form of this stone; the idea, or principle, of this rock type, "hidden" in the primal sandstone matrix, comes to visible presence through the weathering process.

The sculptural character of Hawkesbury Sandstone is evident in its bulbous, hollowed-out formations (see Color Plate IV, p. 135). Elsewhere in Australia this character has been developed further; the sandstone Grampians Range in Victoria has an even more sculptured appearance. In the words of one writer: "Bizarre rock formations abound [in the Grampians]."[229]

Fig. 39. Hawkesbury Sandstone outcrops (charcoal on paper by Nigel Hoffmann)

Other observers, in other parts of the world, have made similar observations concerning sandstone. Victor Hugo writes in his *Les Alpes et les Pyreneés:*

> Sandstone is the most amusing and strangely molded stone. It is among rocks what the elm is among trees. It does not miss a single shape, whim or dream: it has all the appearances and makes all kinds of faces. It seems animated by a multiple soul . . . [230]

The French poet Claudel also writes of sandstone:

> Monstrous stones, fantastically shaped... look like animals from fossil ages... idols whose heads and limbs had been misplaced.[231]

Gaston Bachelard writes that, in the way it evokes for poets visions of forms which are infinitely molded, even to the point of being monstrously distorted, the "sandstone image" is very consistent.[232]

The levity forces at work in the geological organism have moved the convex generative form of intensified (weathered) granite toward concave or degenerated forms; the hollowing-out process is what leads to the individualized sculptural character of weathered sandstone formations. In contrast to the rounded, universal and unindividualized "youthful" character of the granite tor, the weathered sandstone landscape's ancient and highly individual appearance is something like the wizened face of an old person in which we read the language of developed soul quality. Borrowing Hegel's expression, we can call sandstone one expression of the geological organism's "highest subjectivity."[233]

Fig. 40. Comparison of rounded (convex) form of child and hollowed-out (concave) form of old man (from M. Martin [ed.], Educating through Arts and Crafts, Steiner Schools Fellowship Publications, Forest Row, 1999)

The gestures of gravity and levity in the geological organization (Earth — Sun) sound the force field of the octave (see also Chapter 2). The prime sounds the all-potential, undifferentiated force. The major third, in particular, sounds the movement of expansion; the minor third works more inward toward a center and brings about soulfulness. In the development toward the octave, the fourth sounds a degenerating quality; it is the force that turns inwards and consolidates around a center. This degeneration of the expanding-generative forces of the major third as it moves to the fourth is a process of individualization. Sculpturally, the degenerated sphere or concave form speaks of the individualized or differentiated.

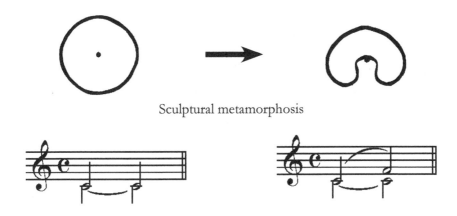

Sculptural metamorphosis

Musical metamorphosis

Fig. 41. Prime to fourth — musical and sculptural metamorphosis of granite to sandstone

Air cognition — the plant organism

The flowering plant is a musical image of Earth-Sun — this much has been discussed in the previous chapter. In the polarity of root and flower the polar gestures of Earth and Sun can be "heard"; in the mediating, spiraling, rhythmical formations of the vegetation the interweaving rhythms of the Earth-Sun relationship come to expression. Building on this primary Inspiration, the task is now to reach toward the specific music of the different plant types in the Yabby Ponds landscape.

Water thinking developed an Imagination of the variety of plants at the Yabby Ponds by "running through" the forms of the different plants along the transect. In the previous chapter it was stated that "the individual changes [in plants and animals] are the various expressions of the archetypal organism that has within itself the ability to take on manifold shapes and that, in any given case, takes on the shape most suited to the surrounding conditions in the outer world." Running through these plants with Water thinking, we came to a preliminary Imagination of the archetypal idea which is sculpting the different plants within this specific landscape. Many different species of plants are here assuming a shape that most suits the surrounding conditions. From this Imagination we distill Inspirations of the formative gesture or musical meaning of the vegetative organization.

The shrubs forming the heathland do not luxuriate in large or expansive rounded leaf shapes; leaf substance does not predominate in the overall structure of the plants. Rather, there is a honed-down or reduced quality; the generative (growth) forces which give rise to rounded leaf forms are held back. Leaves tend toward the linear; they are in many cases spiky and sclerophyllous (xeromorphic). Overall, the plants tend toward a degenerated, concave, sculptured quality. With the larger plants — the twisted and gnarled eucalypts, angophoras and banksias — this sculptured character becomes particularly evident.

Fig. 41. Yabby Ponds — the degenerated, sculptural character of the vegetation

Let us recall the Inspiration from the geological organism: the sandstone has a predominantly sculptural character, resulting from a hollowing out or degenerative process. On the level of gesture we immediately find a correspondence between the geological and plant organizations: the degenerated quality is heightened in the plant organization where it comes to presence in the overall plant morphology. On the level of gesture there is a definite metamorphosis between the geological and plant organizations.

The degenerated, sculptural character of sandstone is what gives it its highly individualized appearance — "it does not miss a single shape." It stands to reason that this power of individualization will be carried into and heightened in the plant organization — and this is indeed the case. Far from presenting a uniform or homogeneous vegetation, this landscape is populated by an enormous number of different plant species, varied in both shape and color.[234] We have seen in the Earth section that the Hawkesbury Sandstone region is one of the floristically richest in Australia, containing some 2,000 indigenous species. In this regard also, it is closely related to the floristically intensely rich sandstone Grampians Ranges in Victoria.[235]

The individualizing power also works within species. Many plants at the Yabby Ponds are sculpted in such a way that each individual tree or shrub is very different from every other. We see this exemplified in the rather gangly eucalypts and banksias in the immediate vicinity of the Yabby Ponds and especially in the stands of angophoras a little further off. This individualized quality can be compared with the uniformity of trees such as the larch, or cypress.

Angophora costata   Cypress

Fig. 42. Comparison of the "individualized" form of an Angophora with the uniformity of a conifer

Now the individualized character of the vegetation can be explored more deeply. Looking at the species along the transect we see that the floral "faces" express a gamut of types. There are "contracted" or "inward" types, such as *Epacris longiflora* with its long tubular floral form (expanding only at the very end). Its superior ovary is enclosed and protected by the petals. Here something is hidden or secreted, confined in a deep inner space. The animals that visit this flower have to be specially equipped to penetrate it to its nectaries: honeyeaters, for example, with their long, curved beaks. These plants could also be called "gravity-types" in that gesturally they belong to the linearity of stem and root; all linearity in plant form expresses the centric geometry of Earth.

Fig. 43. Tubular flower of Epacris longiflora

Polar to this is the expansive gesture of *Hibbertia linearis*, a gesture found in other species in the landscape as well. Its blossom is entirely opened out; the petals fold right back, bringing the center forward. Another plant that takes up the theme of expansion, in an entirely different way, is *Acacia ulicifolia* (prickly moses). In its case it is the stamens that open out in all directions rather than the petals; here, expansion has the quality of directed radiance. These could be called "levity" flower types in that an expansive gesture predominates in them. All gesture of expansion in plant form is related to the all-embracing "lightness" of the Sun.

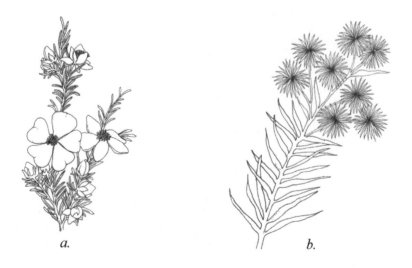

Fig. 44. Comparison of floral forms of: a) Hibbertia linearis; *b)* Acacia ulicifolia

An "in between" plant is *Dillwynia retorta*, a pea; this plant shows a unification of the polarities in the particular way the five petals of the flower form the standard, wings and keel. The flower is opened out or expanded in the standard, and closed or contracted in the keel (which hides the stamens and pistil), with the wings balancing the expanded and contracted gestures.

Fig. 45. Dillwynia retorta *as an "in between" contracted and expanded floral type.*

We can now take the three gestural types and inspiratively gather them between Earth and Sun. Taken separately, each is a one-sided expression of the three-fold principle of the archetypal landscape; each species specializes in one gesture or another, either levity or gravity or some "in between" condition. Viewed together, they present an image of the "wholeness" of the landscape organization. Every species in this landscape could be positioned in a graded sequence between Earth and Sun, which would give an overall picture of gravity-levity in the gestural tendency of this landscape.

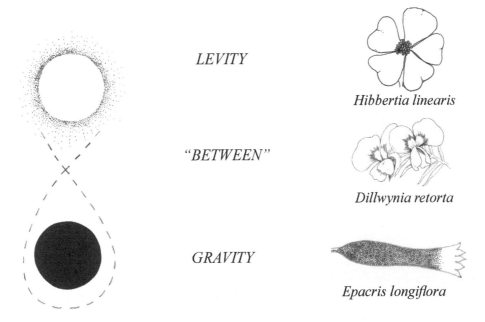

LEVITY

Hibbertia linearis

"BETWEEN"

Dillwynia retorta

GRAVITY

Epacris longiflora

Fig. 46. Yabby Ponds — three floral types as characteristic gestures

The developed Imaginations of *Hibbertia linearis* from the previous section can now be intensified to the point of Inspiration. The expansive tendency of this species belongs not just to the flower, but to the whole plant. Growing in the open spaces of the landscape, *H. linearis* stands erect but not rigidly so, as around a strongly upright central stem. Rather, the stems are thin and long and the habit of the plant is loosely or unevenly erect.

Fig. 47. Hibbertia linearis — *structure of the whole plant*

Likewise the leaves are not spread evenly or densely; leaves appear relatively widely spaced, alternating along the stems in groups of two, with a larger and smaller leaf combined. At the end of the short branches larger groups form with a slightly rounded whorled quality. The vegetative habit of *H. linearis* is loose and opened-out in uneven patterns of expansion and condensation. The plant sprawls or diffuses loosely through space, in a rhythm of linear and rounded gestures.[236] In this respect it can be contrasted with prickly moses (*Acacia ulicifolia*); this plant moves "willfully" as it were, with great directness, and terminates in very sharp points.

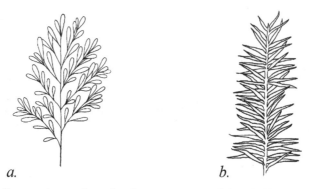

a.   *b.*

Fig. 48. *Comparison of two leaf gestures: a)* Hibbertia linearis*; b)* Acacia
ulicifolia

The leaves of *H. linearis* are linear-oblong in shape and curved down (recurved) at the edges; these qualities together give the leaves an opened-out or spacious character (compared, say, to the rounded cup-like leaves of a hydrangea).

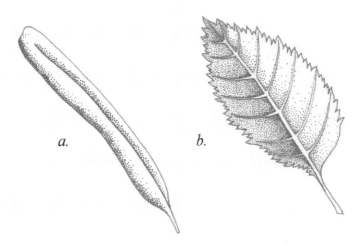

Fig. 49. Comparison of two leaf gestures: a) Hibbertia linearis*; b)* Hydrangea *sp.*

The leaves of *H. linearis* are narrow near the petiole and widen gradually before gathering into a slightly heart-shaped end. The out-stretching gesture develops a gentle expansion but only such that the linear movement is preserved and there is no contraction to a point. The leaf gesture corresponds to the gesture of the plant as a whole: a relatively linear formation condenses into gentle, rounded forms. The musical idea lives in the whole plant, as in the part.

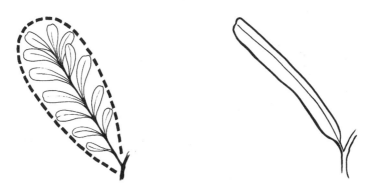

Fig. 50. Hibbertia linearis *— comparison of branch and leaf forms*

The gesture expressed in the vegetative morphology of *H. linearis* is heightened or enhanced in the form of the flower; the flower reveals the idea of expansion and openness which is more or less "hidden" in the more linear vegetative form. We have seen that the flower of this plant is opened out, revealing its center; it is a levity type. The ovary stands above the plane of the petals, and the ovary parts (carpels) are separate from each other (technically, they are "free"). This opened out, free quality is intensified by the profuse sexuality of the many free stamens that crowd around the ovary, and it is amplified by the rich, radiant, melted yellow of the heart-shaped petals which fold back and carry their center before them. This flower is yielding itself, warmly expansive but not willful. The color yellow radiates freely, unlike the focused will force of red (see Color Plate V, p. 135).

Further, the flowers are scattered throughout the vegetative structure; they are not terminal on the stems nor are they clustered. Again, a gesture of focused intent or intense directedness is not evident. The single flowers, rather, are gentle condensations of the spacious vegetative matrix. The idea of freedom, of being opened-out yet condensing into intensities of warm radiance, lives as much in the overall disposition of the flowers as in their individual form (see Color Plate VI, p. 136).

Let us return to the taxonomic Imagination developed in the Water stage through the sequence: order, family, genus, species. We can now intensify this Imagination to Inspiration. A particular levity gesture predominates in the family Dilleniaceae, expressed in the separated carpels and numerous stamens, in the freedom (separateness) of the styles and the petals and the free radiance of the yellow corolla. Dilleniaceae gathers its freedom and force of expansion from universal levity (focused in the Sun). Universal levity is one polarity of the landscape organism; when it marries with its opposite (gravity) the specific organism springs forth, just as colors spring from the marriage of light and darkness.[237] In the formative language of *H. linearis* we "hear" the interwoven gestures of universal levity and gravity.

The different terrestrial landscapes call forth the specific from the universal. The archetypal plant is progressively concretized and made specific through family, order, genus, species — these are moments in a unitary formative movement. The formative music which lives in Dilleniaceae meets the sculpting, individualizing power of the sandstone landscapes of eastern Australia (of which the Yabby Ponds landscape is part). Thus we move toward a developed Intuition of the creation of *H. Linearis*.

Color Plate I. Typical piece of Hawkesbury Sandstone from the Yabby Ponds area. (Pastel on paper by Nigel Hoffmann)

Color Plate II. White-eared honeyeater. (Watercolor on paper by Nigel Hoffmann)

Color Plate III. Four stages in the blossoming and fruiting of Hibbertia linearis. *Water cognition pictures the transitions from one stage to the next. Watercolor on paper by Nigel Hoffmann*

Color Plate IV. Hawkesbury Sandstone outcrop. Acrylic on paper by Nigel Hoffmann

Color Plate V. Blossoming of Hibbertia linearis. *Acrylic on paper by Nigel Hoffmann*

Color Plate VI. Flowering Hibbertia linearis. *Acrylic on paper by Nigel Hoffmann*

Color Plate VII. White-eared honeyeater. Pastel on paper by Nigel Hoffmann

Air cognition — the animal organism

In the language of Inspiration, the animal is the "marriage" of Earth and Sun which has attained to an individuality beyond that of the plant; this much has been explored in the previous chapter. That power of individuality is what we call "soul," and through the animal kingdom soul is heightened and becomes real inwardness; it comes to concrete expression in an animal's form, behavior and cry. With this primary Inspiration in mind, the different animal forms at the Yabby Ponds can now be researched inspiratively.

A first Imagination arose out of our study of the variety of animal types in the Yabby Ponds landscape — the reptiles, marsupials and birds. On the hilltop in the immediate vicinity of the pools, the soul-presence of the animals is one of timidity or reserve. The soul life of the landscape expressed through the animals makes a subdued impression — there is a continuous but quiet twittering of many birds, but the birds themselves only occasionally flit out of the vegetation. One may catch an occasional glimpse of a lace monitor (a reptile) or rock wallaby (a marsupial) by day; the marsupials are mainly nocturnal and solitary. The animals most apparent during daylight hours — the birds — are solitary or in pairs and largely unobtrusive with respect to both their physical size and activity. The animal life here is covert — this is its principal gesture.

If one were to look for the greatest possible contrast, it would be the animals of the African savannas, robustly and overtly asserting their presence through their sheer numbers and size. Something of this robust outward gesture is also found in the buzzing, screeching profusion of insects and birds in the wetlands and coastal lagoons of eastern Australia. Just a quarter of a kilometer distant from the Yabby Ponds, at the base of the hills near the water's edge, the soul atmosphere is somewhat more outward in gesture; here a pair of largish black cockatoos may be observed sitting in a casuarina tree, blithely cracking and eating the nuts; the atmosphere is often filled with squawking sea gulls and the raucous cries of the sulphur -crested cockatoo; large cormorants and other water birds move overtly over the waters in and around the mangroves offshore.

Overtness or outwardness of gesture belongs to animals in which the metabolic-limb system and the soul character of will are primarily developed (see Chapter 3). They carry out their activities in the open; their cry is powerful and directed. The covert, almost timid gesture of the animals around the Yabby Ponds has more of a sense (or sensitive) character (a soul gesture related to the sense-nerve system of animal physiology). Of course, there is great variation within the soul types of the many animals that live in the Yabby Ponds environment. High above may be seen the large and powerful wedge-tailed eagle that, with its piercing shriek, expresses its will character. The white-eared honeyeater is an "in between" type, with a strong will character which alternates rhythmically with a sense quality (or sensitive, timid disposition). A bird with a predominantly sense (or sensitive) gesture is the brown thornbill with its diminutive stature, quiet twitter, and propensity for hiding in the bushes. The size and facial outline alone reveals the relative soul dispositions of these three birds. [238]

Fig. 51. The Yabby Ponds — the contrasting facial gestures of three bird types (from left to right): wedge-tailed eagle; white-eared honeyeater; brown thornbill

Our Imaginations of the white-eared honeyeater can now be intensified into Inspirations; we are reaching toward the specific music of this bird's form and behavior. Physically small, white-eared honeyeaters are mainly solitary but are also seen in pairs, shadowy presences in the vegetation, flitting and perching — momentarily separating from the leaf-greenness, then merging back into it. Most of the white-eared honeyeater's feathers are a dull green/gray; the olive green of its body and gray of the top of its head and hindneck picture its cryptic existence within the dull greens and grays of the vegetation. This is its first principal gesture — an unassertive blending into the foliage, a timidity or reserve.

But in the head region, the black coloration (intensified by the white surround) around the dark beady eye and down into the short curved beak, concentrates and projects forward the forces of the white-eared honeyeater. The intense blackness of its neck and eye, focusing into the curve of the beak, corresponds to its intent darting-out movements as it flits from its perches in the undergrowth to probe for insects and nectar and to assert its territoriality. This is the second principal gesture of its soul life — a quality of focused intent, reaching out and into, penetrating.

In the cry of the white-eared honeyeater the gesture of focused intent attains its most heightened expression. The "toneless tones" of the bird's physical form become the actual tonal music of its cry — it is the same music. Its principal territorial cry is a series of short outbursts which "dart out" of the silence, probe and penetrate the soul space of the landscape before quickly drawing back into the silence. The tones are liquid, mellifluent, flutelike; a sensitivity balances the willfulness of the outburst.

The soul-character of the white-eared honeyeater sounds in the progression of the tones of its principal territorial cry.[239] The cry has three main stages — from the first tone there is a probing upwards through the interval of a third to a high tone of emphasis, before falling away through a third, all in one rapid sliding motion. There are one or two repetitions of this phrase before a silence. In the musical character of the cry — including its punctuation, texture, timbre, and tonal development — the white-eared honeyeater is announcing its soul nature. This is a timidity, a sensitivity, which at the same time seeks to probe, to penetrate and assert. The territory of this animal is not only a physical space but a soul space, determined by the soul-gesture of its cry which sounds forth and announces itself (see Color Plate VII, p. 136).

It is now possible to "hear" the music of the honeyeater in relation to the specific music of the plants which this animal visits; namely, those with floral tubes such as *Epacris longiflora*, the gestural nature of which has already been discussed. The "lightness-sweetness" essence that is being sought by the bird is secreted far into the contracted tubular form of such flowers. We can now inspiratively bring our musical understanding of the bird's form into counterpoint with the music of the *Epacris* and experience their "sounding together." In this way the basic fact of food gathering is intensified to the perception of a drama of soul forces. The bird's entire gesture of sensitive-hiding, then probing-asserting, weaves into the musical dynamic of these flowers. In the first place the honeyeater must come willfully out of hiding to visit these plants. The willful red-yellow of the tubular flowers, standing out at occasional points in open areas of the landscape, "meets" the bird's own willfulness. But the highest longing of the honeyeater's soul is not the assertion aspect but the penetration to find the sweetness essence which the deep centers of the *Epacris* yield. Bird and plant "sound together" in a drama of tension and resolution which forms part of the overall soul-constitution of this landscape organism. This bird sings its own soul nature but at the same time is singing its belonging-together with the hidden levity-spaces of flowers.

As discussed in the previous section, the specific organism is called forth from the universal. In the case of the white-eared honeyeater (*Lichenostomus leucotis*) the creative movement of the archetypal landscape concretizes through the suborder Passeri, family Meliphagidae, genus *Lichenostomus* as it "marries" with the specific formative music of the geological and plant organizations of the Yabby Ponds (and other related landscapes in Australia inhabited by this bird). The Passeri are the "true songbirds," the Meliphagidae the brush-tipped tongued group. Thus we can say that the species *Lichenostomus leucotis* is called into being as a movement within the whole melody of inward, covert soul life of such landscapes, to sing the specific life of soul associated with hidden sweetness and liquidity. In the overall form-language of *Lichenostomus leucotis* we can "hear" the interwoven gestures of gravity and levity, inward reserve and outward assertion, in a music that sounds overtly in its song.

Fire Cognition

The wholeness of the landscape is reflected in every part, every particular form. This study has focused on a particular rock type (Hawkesbury Sandstone), and a plant (*Hibbertia linearis*) and an animal (the white-eared honeyeater) which are two of the many plant and animal species present in the Yabby Ponds environment. In the form-language of these organisms — their physical and musical form distilled and intensified — we read intuitively the creative language of the landscape as a whole.

On many of the hilltops in the region of the Yabby Ponds, in places where the sandstone outcrops are broad and flat, the Garigals — an Australian aboriginal tribe expelled from this landscape not long after the Sydney region was first colonized by Europeans — established sacred sites and marked the rocks with engravings of their gods and spirit beings, often in the form of animals and plants. It is here that they ceremoniously recounted their great creation stories. Through these mytho-poetic intuitions the soul-spiritual character of this landscape was given an explicit artistic form. The task of the Goethean researcher in this landscape is, in a certain sense, to carry forward the work of these original poetic carers of the land. Goethean research also engages artistically with the soul-spiritual form of the landscape but seeks to bring it to consciousness in a scientific manner, in the form of exact Intuitions.

Fire cognition — the geological organism

Granite, the rock of the deep, the siliceous crystal of the Earth,
rises and is worked upon by lightness-forces
which cause the crystalline rigidity to dissolve and dissipate
through Sun-moved winds and waters.
And iron, which belongs to the core, is also taken up within this
levity realm.
It is diffused through all the waters and soft sands
and anchors these sands into sandstone.
Wind and waters work upon this stone of sand
as the sculptor presses and molds from without,
so that the stone moves close to the organic realm:
wizened, bulbous forms, each an individual,
an incipient organism.
Hollows and inner spaces
speak prophetically of a soul life.

Fire cognition — the plant organism

This landscape does not present a verdant, swelling, rounded life,
but rather life turned inward,
sculpted into hard, linear, mineralized leaf formations.
But this curtailment of life allows for a complex soul-revelation
for the great variety of flower forms, the many minute faces —
sentience arises as vegetative life recedes.
This landscape has a multiple soul:
inwardly gathered in the floral tubes of Epacris,
opened out in the radiating, warm sexuality of Hibbertia,
whorled into this flower's exuded gold,
or into the convoluted red flamboyance of Banksia.

Dull, mineralized vegetation
lifts into multiple ecstasies of color and fragrance:
innumerable irruptions of soul.

Fire cognition — the animal organism

Leaf-like bird,
dull green shadow in the barely shifting scrub,
low and sheltered,
hardened greens and ash blacks.
This shadow becomes the bird, which flashes out,
probes and penetrates and claims the light-filled space,
catches the light and liquidity all in a moment,
focuses it and releases it.

Its song is a clear and pure voice of this place,
of its shadows, its silences,
its raw toughness — but also
its hidden sweetness and secret waters.
In this song, this higher flowering,
the landscape becomes a picture of soul.
Over the stream, more fluid still,
a higher liquidity:
pure life of this dry landscape
of raw fire, tough silica and secret pools.

EPILOGUE

Who possesses science and art
Possesses religion as well:
Who does not possess the first two,
O grant him religion.

                          — Goethe[240]

There is a delicate empiricism which makes itself utterly identical with
the object, thereby becoming true theory. But this enhancement of our
mental powers belongs to a highly evolved age.

                          — Goethe[241]

Goethe was one of those creative thinkers whose ideas reach far ahead of their own time. He was fortunate to find favor with a number of prominent scientists and philosophers and he enjoyed being lionized as an artist. But as regards the insights which he felt to be his most precious and significant — in particular those relating to his scientific studies — he knew he was not being understood by most of his contemporaries. Goethe has been accused of being a dilettante, an artist meddling in the sciences; he has been criticized for allowing the artistic to compromise the scientific. Actually it is the permeation of his scientific approach with artistic insight that makes his scientific work so significant for our present age, even if it is at first difficult to make sense of.

In recent times it has become clearer just how progressive and even modern Goethe was as a thinker and an individuality. Henri Bortoft has shown that Goethe's ideas on the history of science really predicted the postmodern way of thinking about science, in particular his view that science is culturally or historically embedded.[242] If we look at Goethe's writings it is not hard to find evidence of the liberalism and pluralism which give his work a modern feel. For example, he lauds New York (of the early 1800s) for being a city where ninety different Christian sects could exist without interfering with each other. "In scientific research — indeed, in any kind of research," he goes on to say, "we need to reach this goal..."[243] In another place he writes that "we cannot find enough points of view nor develop in ourselves enough organs of perception to avoid killing [the organic being] when we analyze it."[244] What comes forward here is Goethe's desire to break out of traditional molds and dogmatic structures that lead to fixed attitudes and one-sided methods of understanding the world. A connection could be made to twentieth-century postmodern thinking in which truths of any kind — scientific, religious, philosophical — become something of the past, where more interest is taken in the variety of "subjective perspectives."

There is no question that Goethe's thinking has prophetic overtones of modernity and postmodernity. But it would be a great mistake — in terms of a right understanding of his way of science — to assume that the views of the twentieth century had overtaken and rendered Goethe's approach obsolete. In truth, Goethe's way of science still belongs to our future, even here at the beginning of the third millennium. He knew that the truth of color, of organic form, could only be approached by letting the phenomena "speak" on their own terms, not in terms of a system of thought imposed on them. This is the "delicate empiricism" which he considered to belong to a highly evolved age. For this delicate empiricism to become a reality as a research methodology there needs to be an extraordinary openness and flexibility of attitude on the part of the researcher. Even the humblest weed or rock of the field, Goethe saw, holds mysteries that can call to us if we are disposed to "hear" them. In an attitude of reverence the human being has the possibility of receiving the gifts of nature's "open secrets." This was the real reason for Goethe's pluralism and liberalism — not to do away with truth, but to find a higher truth.

When Goethe writes that every approach we make in the study of nature should spring from "all [our] united powers," he is likewise indicating that the researcher needs to open up every possible mode of perception, and here he means both the scientific and the artistic ways of understanding, the logical and the imaginative mind.[245] From this it is clear that pluralism, for him, means nothing in and of itself unless it is governed by a unifying principle. Simply taking many and varied points of view is something which he thought of as fragmentary and "blameworthy" because it can never penetrate to the wholeness of the phenomenon. Goethe, in his research into living form, sought the unity of the artistic and the scientific, not a sum total of artistic and scientific perspectives — as if these could add up to a whole.[246] His way of seeing springs from the actual marriage of the scientific and the artistic, a union in which both become something different, in and through each other.

If we realize a true unity of science and art in our research process, then we also have religion — that is Goethe's bold claim. Here we have a suggestion, a pointer, to what is most prophetic in his approach, to what still belongs to our future. In this respect it is not particularly helpful just to call Goethe a "Renaissance man" and categorize him with Leonardo as a unique, multi-talented historical personage. Goethe knew that science and art must find each other again to achieve a true science of living form, and he envisioned this science-in-potential as permeated with a religious mood. The implication is that, through finding the unity of science and art in our research methodology, every human being has the possibility of discovering — or rediscovering — his or her religious nature. This might strike us as somewhat strange, retrogressive and perhaps, after all, very "unmodern." Actually Goethe means religion in a particular sense when he refers to the unity of science and art. He is referring to the religious aspect of the human soul, the religious way of understanding the world. He means the form of perception which can actually see the working of the spiritual in nature, which grasps in utmost clarity that nature is a manifestation of the spiritual or archetypal. In these terms it is quite accurate to speak of Goethe's way of research as being on the threshold of a science of the spiritual.

Everything in the foregoing chapters is intended as an illustration of how Goethe's method opens up a way of perceiving the spiritual in nature. Much else could be said as to the way this science leads to new approaches in architecture, agriculture, medicine, education, and other spheres of the sciences and arts. The question of education is actually central because growth in all other areas of research and practice will depend on the development of effective ways of teaching Goethe's way of science. Beginnings have already been made in relation to both elementary and high school education.[247] How the modern university could be transformed so that it can provide an adequate context for the teaching of Goethe's way of science is a large and difficult question, but one that will need to be addressed.

Undoubtedly many people in the contemporary world still feel as the writer D. H. Lawrence did — that a true science of living form is, as yet, "quite closed to us" (see p. 3). One can have an instinctive sense that the purely logical, quantitative or mechanistic approach to nature is inappropriate or one-sided, an imposition on the living being of nature, even a destructive force, and still have no inkling of the fact that a different kind of science actually exists. Certainly it remains no easy matter to discover alternative or complementary scientific approaches when these form no part of conventional education at any level. Those maturing in their academic studies, young adults working their way through higher education or moving out to work in the diverse areas of society, those who are still testing the waters of what modern society has to offer — such individuals may easily come to the conclusion that no viable alternatives exist for them.

Goethe had trouble understanding why many of his contemporaries could not understand his ideas and methods, and it has to be said that it is still possible they will not be understood today. But ideas ripen — and, as has been indicated in this book, a whole approach, a methodology, has developed on the basis of Goethe's work. It has been the goal of this book to make this methodology more readily comprehensible and more accessible so that it can be taken up and applied in all walks of life. Goethe's science of living form has the potential to meet a real need of the present world: the need to be able to apprehend and work with the creative, spiritual dimensions of nature.

NOTES

CHAPTER 1

1. D.H. Lawrence, *Fantasia of the Unconscious*, (London: Heinmann, 1961), p. 6.

2. Rudolf Steiner, *Therapeutic Insights, Earthly and Cosmic Laws*, (Spring Valley: Mercury Press, 1984), p. 17.

3. J.W. von Goethe, from his autobiography, quoted in H.B. Nisbet, *Goethe and the Scientific Tradition*, (London: University of London, 1972), p. 68.

4. A.J. Husemann, *The Harmony of the Human Body*, (Edinburgh: Floris Books, 1994), p. 235.

5. " ...the frenziedness of technology may entrench itself everywhere to such an extent that someday, throughout everything technological, the essence of technology may come to presence in the coming-to-pass of truth. Because the essence of technology is nothing technological, essential reflection upon technology and decisive confrontation with it must happen in a realm that is, on the one hand, akin to the essence of technology and, on the other, fundamentally different from it. Such a realm is art." M. Heidegger, *The Question Concerning Technology and other Essays*, (Grand Rapids: Harper, 1977), p. 35.

6. In the first half of the eighteenth century the writer Julien La Mettrie had published the books *Man a Machine* and *The System of Nature* that were widely circulated in Europe.

7. Goethe, *Scientific Studies*, (New York: Suhrkamp Publishers, 1988), p. 46.

8. R. Steiner, *Goethean Science*, (Spring Valley, NY: Mercury Press, 1988), p. 76.

9. Arthur Zajonc, "Goethe and the Science of His Time," in *Goethe's Way of Science: A Phenomenology of Nature*, (ed. by D. Seamon & A. Zajonc), (New York: SUNY, 1998), p. 21.

10. Goethe was the founder of the science of morphology, and his scientific writings amount to many volumes. In the 1880s Steiner, as a young scholar, was given the task of editing Goethe's scientific writings,

which had lain for decades in the Goethe archives in Weimar. The book *Goethean Science,* recently retranslated as *Nature's Open Secret,* comprises Steiner's introductions to the first major edition of Goethe's scientific works in the German language (Kürschner's *Deutsche National Literatur,* in four volumes). Steiner went on to further develop and apply Goethean science in the areas of medicine, agriculture, education, architecture, therapy, the arts and economics, among others.

11.   R. Steiner, *A Modern Art of Education,* (London: Rudolf Steiner Press, 1954), p. 33.

12.   R. Steiner, *Therapeutic Insights, Earthly and Cosmic Laws,* pp. 17-18. Rudolf Steiner discusses how Goethe, in his study of plant growth and metamorphosis, attained to the level of Imagination but stopped short of the levels of Inspiration and Intuition. In this respect, Steiner goes far beyond Goethe in his study of organic form. Steiner shows, specifically, that Inspiration is necessary to understand animal organization, and Intuition to understand the human being. R. Steiner, *Fruits of Anthroposophy,* (London: Rudolf Steiner Press, 1986), pp. 36-45.

13.   See Frederick Amrine, "The Metamorphosis of the Scientist," in *Goethe's Way of Science: A Phenomenology of Nature,* pp. 34-35. The requirement to produce a new hypothesis is part of the scientific method, but the productive/creative activity itself is unstructured and without a methodological basis.

14.   Goethe, *Scientific Studies,* pp. 24-25.

15.   Ibid., pp. 74-75.

16.   Ibid., p. 74.

17.   Ibid., p. 13.

18.   Fritz Heinmann, "Goethe's Phenomenological Method," in *Philosophy* 9 (1934), p. 73.

19.   Husserl's method proceeds by what he calls a "phenomenological reduction" in three stages. Imagination is that intentional act which, in the first place, frees ("neutralizes") consciousness from the factual world as given through perception (first phase of reduction). Husserl does not mean that this world is denied or rendered fictitious in the sense of being understood as the *mere* product of the human imagination — only that thinking is released from the habitual view that the perceived world is "the real" to which ideas must somehow find a correspondence. Consciousness grasps that it itself is *intending* the world of perceived entities and that, in as much as entities appear as standing over and against us, it is intending or seeing the world in one particular way. Husserl calls this the "natural attitude" or "habitual standpoint," and it is here that we find a correspondence with Goethe's notion of the "empirical phenomenon" — "what everyone finds in nature."

Through the stages of phenomenological reduction, the imagination comes to experience itself in the realm of free possibility, grasping the entities of sense perception not so much as fixed facts but as possibilities or potentials (second stage of reduction). Husserl designates this "free variation" or "ideation" and regards this activity of the imagination, paradoxically, as the very means by which the mind grasps the necessary laws (*telos*) which bring particular phenomena into being. At this point we find a correspondence with Goethe's notion of the "scientific phenomenon" — "producing [the phenomenon] under circumstances and conditions different from those in which it was first observed."

Human imagination does not make or construct the laws of nature arbitrarily but *discovers* them. By virtue of the freedom of imagination, a person is able to apprehend the particular coherence of meaning which allows them to say what a particular thing *is*. Consciousness, paradoxically, both *apprehends* or "*sees*" (s*chauen*) and *constitutes* the world, according to Husserl (third stage of reduction). Here both Goethe and Husserl speak of the "pure phenomenon" or "pure idea" realized through intuitive understanding; we find here a similarity in their description of a movement from the objectively *constituted* to the free, creative, productive or *constituting* power of mind.

20. Martin Heidegger, *Being and Time*, (trans. by J. Macquarrie & E. Robinson), (Oxford: Blackwell, 1992), p. 61.

21. Goethe, *Scientific Studies*, p. 22.

22. Ibid., p. 11.

23. Quoted in Viktor Zuckerkandl, *Sound and Symbol*, (New York: Princeton University Press, 1956), p. ix.

24. It is not hard to understand why Goethe was so attracted to the educational ideas of Rousseau, for this thinker was saying something similar with respect to the education of the child. Our way of education, Rousseau insists, should not be foisted on the child but should belong to the actual nature of the child; as a methodology it should be effected through loving concern for the unfolding human individuality.

25. When Goethe asked: "What higher synthesis is there than a living organism," he was referring to the irreducible wholeness of living entities (Goethe, *Scientific Studies*, p. 49). See also the discussion in R. Steiner, *The Science of Knowing*, (New York: Mercury Press, 1988), p. 86.

26. See M.E. Zimmerman, *Heidegger's Confrontation with Modernity*, (Bloomington: Indiana University Press, 1990), p. 222.

27. Henri Bortoft, *The Wholeness of Nature*, (Hudson: Lindisfarne Press, 1996), p. 369.

28. The founder of systems theory, Ludwig von Bertalanffy, asserts that his theory is "an offspring of Goethe's conception" (L. von Bertalanffy,

"Goethe's Concept of Nature," *Main Currents* 8 [1951], p. 82). Two of the principal originators of chaos theory — Mitchell Feigenbaum and Albert Libchaber — were strongly influenced by Goethe. Feigenbaum considered Goethe's "holistic" theory of color to be correct. Libchaber was inspired by Goethe's monograph *On the Transformation of Plants*, as well as the work of the Goethean scientist Theodor Schwenk on the creation of flowing forms in water and air (see James Gleick, *Chaos: Making a New Science*, [New York: Viking, 1987], pp. 165 & 197). Brian Goodwin writes that his application of complexity theory to the study of organic form is "very much in the Goethean spirit" (B.C. Goodwin, *How the Leopard Changed Its Spots*, [London: Phoenix, 1994], p. 123).

29. Such are precisely the grave concerns expressed by Jeremy Rifkin in his *Algeny*, that the new understanding of molecular genetic functions combined with developments in the sciences of complexity and morphogenetic fields, is making huge advances toward a culture founded upon the creation of engineered life forms, the "Age of Algeny" as he names it. This biotechnological age, as he sees it, will be ushered in in three distinct phases: the first is already well under way — the understanding of how to modify genes and insert them into organisms. The second phase depends upon an understanding of the cybernetic relationships between gene, cell and organism and relates to new developments in systems theory, process morphology, complexity science and the like. Now, rather than merely modifying particular characters, entire organisms may be engineered. The third phase is the engineering of entire ecosystems. In the Age of Algeny talk of "creative possibilities," "useful models," or "perspectives" will gradually replace notions of "permanent" or "objective" truth. J. Rifkin, *Algeny*, (New York: Viking Books, 1983), pp. 215-243.

30. B.C. Goodwin, *How the Leopard Changed Its Spots*, p. 103.

31. Ibid., p. 82.

32. Goodwin (ibid.) describes how, for Goethe, the investigation of nature involved "an aesthetic appreciation of form and quality quite as much as dynamic regularity" (p. 123). However, after discussing how Goethe saw organisms as "dynamic forms in transformation," Goodwin goes on to say that only in the twentieth century have the mathematical tools been developed to actually carry out the analysis of dynamic form that Goethe had recognized (i.e., through his aesthetic appreciation).

33. See Martin Heidegger, "The Origin of the Work of Art," in *Poetry, Language, Thought*, (New York: Harper Colophon Books, 1971).

34. Aristotle declared: "Art acts like nature in producing things" which, Jane Harrison notes, has long been mistranslated to read "art imitates nature." J. Harrison, *Ancient Art and Ritual*, (London: Moonraker, 1978), pp. 108-9.

35. See S. Waterlow, *Nature, Change and Agency in Aristotle's Physics*, (Oxford: Clarendon, 1982), pp. 1-45.

36. Imanuel Kant, *Critique of Judgment*, (New York: Hafner, 1951), p. 220.

37. Frederick von Schelling, *System of Transcendental Idealism*, (trans. by Michael Vater), (Charlotteville: University Press of Virginia, 1978), p. 126.

38. This question is discussed in Chapter 3 of this book, in the section "Evolution as Creative Process." Here it is shown how a non-mechanistic, living form of understanding, which recognizes the creative nature of the organism, its "inner completeness," comprehends how an organism *shapes itself* by responding creatively to external environmental conditions.

39. The full quotation is: "We can . . . think of an understanding which being, not like ours, discursive, but intuitive, proceeds from the *synthetical-universal* (the intuition of a whole as such) to the particular, i.e., from the whole to the parts. . . It is here not at all requisite to prove that . . . an *intellectus archetypus* is possible, but only that we are led to the idea of it — which too contains no contradiction — in contrast to our discursive understanding, which has need of images (*intellectus ectypus*) and to the contingency of its constitution." Kant, *Critique of Judgement*, p. 289. Goethe himself made comment on this passage (see Goethe, *Scientific Studies*, p. 31).

40. Michael Vater, introduction to Schelling, *System of Transcendental Idealism*, p. xxiv.

41. Craig Holdrege discusses the notion of the gene as the fundamental cause or atom of the organism as an example of "misplaced concreteness" in *Genetics and the Manipulation of Life: The Forgotten Factor of Context*, (Hudson: Lindisfarne Press, 1996), pp. 86-90.

42. It is true that artworks have an external artificer: the artist. However, artworks are not machines and artists are not technicians. The artist is an agent who allows the artwork to take form organically. Goethe understood this and spoke of the artwork as a "spiritual creation" [Goethe, *Conversations with Eckermann*, (Washington DC; M. Walter Dunne, 1901), p. 377.] Elsewhere he spoke of classical Greek sculpture as "works of Nature, brought forth by human beings in accordance with true and natural laws." Goethe, *Italian Journey*, Sept. 6, 1787.

43. See E.-M. Kranich, *Thinking Beyond Darwin*, (Hudson: Lindisfarne Books, 1999), p. 143.

44. An essay that connects Polanyi's insights with Goethe's way of science is *The Third Culture—Participatory Science as the Basis of a Healing Culture* by J. M. Barnes (Hillsdale, N.Y.: Adonis Press, 2009.) It must be borne in mind that animal nature has sentience as its distinguishing feature, but also growth (it carries within its nature the plant). Likewise with the human being: personhood is its distinguishing feature, but it also embraces sentience and growth (it carries within its nature the animal and plant).

45. See F. Amrine, "The Metamorphosis of the Scientist," in *Goethe's Way of Science: A Phenomenology of Nature*, pp. 38-39.

46. See H. Bortoft, *The Wholeness of Nature*, p. 280.

47. D. Hyland, *The Origins of Philosophy*, (New York: Capricorn, 1973), p. 254.

48. Aristotle, *De Anima*, (Harmondsworth: Penguin Books, 1986), p. 137.

49. See Keith Critchlow, *The Soul as Sphere & Androgyne*, (Ipswich, Golganooza Press, 1980). Empedocles considered that the whole of universal life works as a dynamic of Love and Strife; Earth is Strife (difference) and its polarity is Love (unity), and their interaction is conceived of in the form of a circle. Here we have an imaginative picturing of the principle of organism or wholeness, the unchanging One which is creating the many.

50. M. Tiles, *Bachelard: Science and Objectivity*, (Cambridge: Cambridge University Press, 1984), pp. 53-58.

51. Collins Concise Dictionary, (London: Collins, 1988), p. 361.

52. See Gaston Bachelard's works *The Psychoanalysis of Fire, Water and Dreams — An Essay on the Imagination of Matter; Air and Dreams — An Essay on the Imagination of Movement; Earth and Reveries of Will; Earth and Reveries of Repose; The Flame of the Candle* and *The Fragment of a Poetics of Fire*.

53. G. Bachelard, *On Poetic Imagination and Reverie*, (Indianapolis: Bobbs-Merrill Press, 1971), p. xiv.

54. G. Bachelard, *Air and Dreams*, (Dallas: The Dallas Institute, 1988), p. 11.

55. G. Bachelard, *On Poetic Imagination and Reverie*, pp. 11-12.

56. G. Bachelard, *Water and Dreams*, (Dallas: The Pegasus Foundation, 1983), p. 16.

57. As Colette Gaudin observes: "At times [Bachelard] seems very close to Novalis's idea that man is the speaker for the poetry of nature: 'The world imagines itself in human reverie.' But in fact, Bachelard leaves the question open: "When a dreamer speaks, who speaks, he or the world?" (Bachelard, *On Poetic Imagination and Reverie*, xxv).

58. G. Bachelard, *On Poetic Imagination and Reverie*, p. 7.

59. A discussion of the Elements is central to Hegel's natural philosophy in which he endeavors to show how life (which he calls the "universal individual") makes itself concrete through the different kingdoms of the natural world. Certainly the Elements do not figure within Hegel's philosophy as a central preoccupation; to the extent that they do, they advance the modern philosophy of nature in a very significant way. Thinking on the Elements forms part of Hegel's *Philosophy of Nature* which itself is the second part of his *Encyclopaedia of the Philosophical Sciences*, under the heading of "Physics," and then again in the third section, "Organics," where he connects three of the Elements to the three traditional kingdoms of nature — Earth with the mineral kingdom, Water with the plant kingdom and Fire with the animal kingdom. Hegel calls the Elements "universal basic physical forms" and "universal natural existences." In the context of his discussion, Hegel makes this observation: "From the standpoint of chemistry, we are, it is held, required to understand by an 'element' a general constituent of

bodies which are all supposed to consist of a definite number of these elements. Men start by assuming that all bodies are composite, and the concern of thought is then to reduce the infinite variety of qualified and individualized corporealities to a few simple incomposite and therefore general qualities. Based on this supposition, the conception of the four Elements which has been general since the time of Empedocles, is nowadays rejected as a childish belief because, forsooth, the Elements are composite. No physicist or chemist, in fact no educated person, is any longer permitted to mention the four Elements anywhere."

60. See Jochen Bockemühl, "Elements and Ethers: Modes of Observing the World," in *Toward a Phenomenology of the Etheric World*, (New York: Anthroposophic Press, 1985), pp. 1-67, and Georg Maier, "Die Elemente als Stufen der Naturbetrachtung," [The elements as stages of nature observation], *Elemente der Naturwissenschaft* 13 (1970), pp. 1-9.

61. R. Steiner, *Therapeutic Insights, Earthly and Cosmic Laws*, pp. 17-18.

62. Steiner goes to considerable trouble to clarify this point. In one place he writes: "One can form a general concept of 'mysticism.' According to it, mysticism comprises what one can experience of the world through inner soul experience. This concept, first of all, cannot be disputed. For there is such an experience. And it reveals not only something about man's inner being but also something about the world . . . But one must bring the full clarity of concepts into the experiences of the mystical organ if knowledge is to arise. There are people, however, who wish to take refuge in what is "inward" in order to flee the clarity of concepts . . . My writings everywhere speak *against* this mysticism; every page of my books, however, was written *for* the mysticism that holds fast to the clarity of ideas in thinking and that makes into a soul organ of perception that mystical sense which is active in the same region of man's being where otherwise dim feelings hold sway. This sense is for the spiritual completely like what the eye or ear is for the physical" (R. Steiner, *The Science of Knowing*, pp. 127-128).

CHAPTER 2

63. R.G. Collingwood, *The Idea of Nature*, (Oxford: Clarendon Press, 1945), p. 126.

64. Earth was one of the four Elements in the characterization of Aristotle in his *De Anima*, this having been drawn from a more ancient tradition. In his Theogony, the ancient Greek poet Hesiod presents an account of the goddess Gaia (Earth) separating from Ouranos (Sky) by an act of violent dismemberment; this was also described by the poet Euripides.

Earth, for the thinkers of old, named the definite, solid aspect of the world-soul which we now call "matter" (the conditioned or determined). Earth is the realm of the actual, the manifest or formed, as opposed to the active, forming power of Fire (potency, or the creative Word). Earth names the realm of definite, rigid substance, highly condensed and particulate, ultimately the atoms of Democritus, the realm of separation and separateness. It names the physical or embodied nature of things in general — seen as dark, lost, in a state of strife and without wisdom, fallen and consigned to the shadowy realm of Plato's Cave. The pre-Socratic sage Parmenides, in his *Proem*, depicted the journey out of the darkness of the Earth (opinion) into the light of the Sun (wisdom). Hesiod poetically evoked the association of Earth with Hades, the dark, underground realm of the dead (see, for example, G.S. Kirk & J.E. Raven, *The Presocratic Philosophers*, (Cambridge: Cambridge University Press, 1957); D.E. Hahm, *The Origins of Stoic Cosmology*, (Columbus: Ohio State University Press, 1977); D. Hyland, *The Origins of Philosophy*, (N.Y.: Capricorn Books, 1973); and F.M. Cornford, *From Religion to Philosophy*, (N.Y.: Harper & Row, 1957).

65. Gaston Bachelard, *Earth and Reveries: An Essay on the Imagination of Forces* in *Transcendental Dynamics: A Bachelardian Romantic Perspective*, PhD Dissertation by L. Zancu, (Ann Arbor: University Microfilms, 1975), p. 91.

66. G. Bachelard, *Earth and Reveries*, p. 95.

67. Those relativist philosophies of science which seek to deny the possibility of objective knowledge, which are "celebrations of subjectivity," are still expressions of Earth thinking. Just because attention is turned toward such things as the social conditioning of the mind does not mean that anything other than the "solid" mode of thinking is being applied. Here the knower (the subject) stands as an observer external to the objects of knowledge which in this case are psychological factors such as the history of a person's familial and social conditioning or their personal disposition and temperament. In other words, the opposition of subject and object itself arises from the experience of externality or otherness. As Heidegger writes, everything "anti" necessarily remains "held fast in the essence of that over against which it moves." Martin Heidegger, *The Question Concerning Technology and Other Essays*, p. 61.

68. H. Bortoft, *The Wholeness of Nature*, p. 176.

69. Ibid., p. 392.

70. An example of scientific realism is the account given of the existence and properties of atoms, molecules and sub-atomic particles in the standard chemistry textbook. Even though these phenomena are not sense-perceptible, they are held to be "strongly objective" because they are deduced by conventional scientific methods.

71. R. Trigg, *Reality at Risk*, (Brighton: Harvester, 1980), p. 83.

72. The epistemological problem arises when thinking tries to ascertain whether it can *really* form a cognitive relationship with external reality. Thinking occupies itself with this problem in terms of the traditional notion of truth as correspondence, rather than by reflection upon the ways in which this notion of truth might *itself* be limited. Heidegger has pointed out that this understanding of truth was merely assumed by Kant and had no reason to be (Martin Heidegger, *Being and Time*, p. 258). Rudolf Steiner writes: "According to [the customary] concept, knowing is supposed to consist in making a copy of the real conditions that stand outside our consciousness and exist *in-themselves*. But one will be able to make nothing out of the possibility of knowledge until one has answered the question as to the what of knowing itself. The question: *What is knowing?* thereby becomes the primary one for epistemology" (Steiner, *Goethean Science*, p. 106).

73. It is true that realists do toy with the idea that qualities may be real and not reducible to quantities or that reality may include "unmeasurable quantities" (Trigg, *Reality at Risk*, p. 169). These are philosophical speculations; how such qualities are established in scientific procedure is another matter. To all intents and purposes, science establishes itself quantitatively.

74. Martin Heidegger, *The Question Concerning Technology and Other Essays*, p. 65.

75. Henri Bergson, *Creative Evolution*, (London: Macmillan,1960), p. x.

76. See J. Lear, *Aristotle: The Desire to Understand*, (Cambridge: Cambridge University Press, 1988), pp. 231-246. When a later mathematical science such as biological taxonomy tried to take into account qualities of organisms such as their color, taste and scent, it could only do so by turning these qualities into quantities — that is, into terms of presence or absence, 1 or 0 — thus rendering them mathematically workable.

77. Precisely this gesture of thinking has led to the search for the atom, the supposedly fundamental unit of matter. Before techniques were available which could penetrate and "externalize" the hidden recesses of matter it was the ideal of physical science to find the atom; when such techniques became available it was realized that the atom was not fundamental at all, and the search continued to yield more and more "fundamental particles," which themselves turned out to not be fundamental.

78. See Steiner, *The Science of Knowing*, p. 90.

79. Bergson, *Creative Evolution*, p. ix.

80. J. Hirschberger, *A Short History of Western Philosophy*, (Guildford: Lutterworth Press, 1976), p. 28.

81. See Oleg Polunin, *Flowering Plants of Europe: A Field Guide*, (London: Oxford University Press, 1969), pp. 117-123.

82. G.H.M. Lawrence, *Taxonomy of Vascular Plants*, (N.Y.: Macmillan, 1951), p. 520-521; A. Cronquist, *An Integrated System of Classification of Flowering Plants*, (N.Y.: Columbia University Press, 1981), p. 446 and G. Bateman, ed.), *Flowering Plants of the World*, (London: B.T. Batsford, 1978), pp. 119-122. For Goethean studies of this family see G. Grohmann, *The Plant* (Vol. II), (London: Rudolf Steiner Press, 1974), pp. 65-78; and W. Pelikan, "The Cruciferae," *The British Homoeopathic Journal*, Vol. LXIV, No. 1 (Jan. 1975), pp. 57-64, and No. 2. (April 1975), pp. 107-112.

83. That is, shaped like a lyre: having a large terminal lobe and smaller lateral lobes.

84. The pseudo-violet *Ionopsidium acaule*, from Portugal, germinates within three days and opens its first flowers after a fortnight; many other crucifers show the same tendency.

85. When mustard seeds are chewed, for example, the glucosinolate (sinigrin) in them is split into volatile allyl isothiocyanate (mustard oil glucoside), dextrose (a sugar) and potassium bisulphate (a mineral salt). W. Pelikan, "The Cruciferae," *The British Homoeopathic Journal*, Vol. LXIV, No. 1, p. 61.

86. Goethe, *Scientific Studies*, p. 11.

87. See Martin Heidegger's discussion on the nature of reflection in science in *The Question Concerning Technology and Other Essays*, pp. 155-182.

88. We find the mechanical image coming to the fore in contemporary culture as a way of thinking about the world. See E. J. Dijksterhuis, *The Mechanization of the World Picture*, (Oxford: Clarendon Press, 1961).

89. H. Bergson, *Creative Evolution*, p. x.

90. Ibid., p. 203.

91. The pre-Socratic philosopher Thales is probably best known in connection with the Element Water, which he believed to constitute the soul and be the "first principle" of all things. Thales is reported to have said: "*Nous* [mind] is quickest of all, for it runs through everything." The Greek Orphic traditions conceived of a "world egg" which arose out of water and slime and which was bounded about by the primal river Okeanos, as by a serpent. Similarly, the cosmic whole was later depicted as a serpent with its tail in its mouth: Ourobouros, a symbol of the "unbounded" and "continuous." The serpent-river was associated with the primal cosmic or procreative power, a life-fluid or semen. See G.S. Kirk & J.E. Raven, *The Presocratic Philosophers*; D.E. Hahm, *The Origins of Stoic Cosmology*; D. Hyland, *The Origins of Philosophy;* and F.M. Cornford, *From Religion to Philosophy*.

92. See Theodor Schwenk, *Sensitive Chaos*, (London: Rudolf Steiner Press, 1965).

93. G. Bachelard, *Water and Dreams*, p. 108.

94. Ibid., p. 108.

95. Ibid., p. 107.

96. For intellectual, analytical thought, time is a function of the movement of bodies through space.

97. Goethe, *Scientific Studies*, p. 64.

98. J. Bockemühl, *Toward a Phenomenology of the Etheric World*, p. 12.

99. Ibid., p. 11.

100. Michael Polanyi, *Meaning*, (Chicago: The University of Chicago Press, 1975), p. 36. Polanyi also writes: "We ... share the purposes or functions of any living matter by dwelling in its motions in our efforts to understand their meaning" (ibid., p. 45).

101. In some plants all these parts are visible simultaneously on the plant — for example, flower and fruit — in other plants not. However, over the time of the plant's growth, empirical data can be gathered and the growth sequence understood.

102. See R. Steiner, *The Science of Knowing*, p. 97.

103. For a discussion of the reasons why Goethe saw his method of studying the qualities of color as mathematical, and his criticism of Newton's quantitative method as being mathematically defective, see Bortoft, *The Wholeness of Nature*, p. 229.

104. R. Brady, "Goethe's morphology" in *Goethe and the Sciences: A Reappraisal*, p. 297.

105. Ibid., p. 278.

106. Margaret Colquhoun presents the different forms of cabbage as a "Goethean" exercise in "exact sensory imagination" leading to an "archetypal experience." She writes: "Brussels sprouts you will know are swollen side buds. Try contracting these and pulling down the stem, buds and leaves making a tight ball — a cabbage! Or let the leaves expand and grow round and crinkly into kale. The stem swelling (reduce all buds) produces kohlrabi, and if the inflorescence is allowed to swell we find cauliflowers . . . calabrese or broccoli. Below the ground, swelling radishes are a further step and, at the top, oil seed rape and mustard show the hot sulphur smell and taste within the seeds. All these plants are closely related; all show something of the sulphur hot taste and smell of cabbage." M. Colquhoun & A. Ewald, *New Eyes for Plants*, (Stroud: Hawthorn Press, 1996), p. 179.

107. J. Lear, *Aristotle: The Desire to Understand*, p. 19.

108. G. Bachelard, *Air and Dreams*, p. 163.

109. H. Bortoft, *The Wholeness of Nature*, p. 242. Italics added by the author.

110. Ibid., p. 242.

111. Ibid., p. 382.

112. The Element Air is usually associated with Anaximenes who considered it to be the first principle (*arche*) of all things, boundless and all-embracing, giving rise to Fire through the process of rarefaction and to Water and Earth through the process of condensation. It is the aspect of world soul which expresses itself as psyche; it is mobile, transparent (when pure), all-pervading. In the religious poetry of the Orphics "the soul enters into us from the whole as we breathe, borne by the winds" (Gk. *pneuma* means breath or spirit). Theophrastus wrote that "the air within us is a small portion of the god" and that Air "holds us together." In its most rarefied aspect it was called the aether, the blazing clarity of the heavens, not so much a physical region as the atmospheric clarity "which makes things visible"; an active force which "breaks through" from the heavens and allows "every crag and glen of the mountain to be seen." C.H. Kahn, *Anaximander and the Origins of Greek Cosmology*, (N.Y.: Columbia University Press, 1960), p. 142. See also G.S. Kirk & J.E. Raven, *The Presocratic Philosophers*; D.E. Hahm, *The Origins of Stoic Cosmology*; D. Hyland, *The Origins of Philosophy;* and F.M. Cornford, *From Religion to Philosophy*.

113. J. Bockemühl, *Toward a Phenomenology of the Etheric World*, p. 26.

114. G. J. Seidel, *Martin Heidegger and the Pre-Socratics*, (Lincoln: University of Nebraska Press, 1964), p. 64.

115. Aristotle, *De Anima*, p. 219. For an explanation of this passage in terms of heightened thinking, see J. Lear, *Aristotle: The Desire to Understand*, p. 139.

116. This is why the nature of music has for so long baffled thinkers, many of whom arrived at the conclusion that it is merely the medium for subjective outpouring. The philosopher Susanne Langer has done a great deal to correct this notion; as she shows, far from being a mere expression of feelings, music represents a definite experience and understanding of the world. She argues that the musical sense experiences meaning and articulates it through a non-discursive language, which she calls a "presentational" symbolic language. She writes: "A composer not only indicates, but *articulates* subtle complexes of feeling that language cannot even name, let alone set forth; he knows the forms of emotion and can handle them, "compose them" and "the lasting effect of [music] is . . . *to make things conceivable* rather than to store up propositions. Not communication but insight is the gift of music . . . " Susanne Langer, *Philosophy in a New Key*, (New York: Mentor Books, 1951), pp. 188-207. While Langer goes a long way towards showing that music is not merely the expression of feelings, her discussion limits music's domain to the human psyche, as

if it has no connection to external phenomena. It is Victor Zuckerkandl who has made it clear that music is dealing with the "inner life" of nature altogether, with the forces which work in living form just as much as in the human psyche.

117. Viktor Zuckerkandl, *Man the Musician*, (New York: Princeton University Press, 1973), p. 98.

118. The octave, according to Zuckerkandl, is experienced as a "returning-to-the same," *Sound and Symbol*, pp. 88-100. The tone of the octave is both the same and different; it is precisely the same note as the prime, yet at a higher pitch. The One releases itself into the multiplicity of the scale and finds itself again in the octave. See Dane Rudhyar, *The Magic of Tone and the Art of Music*, (Boulder, CO: Shambhala, 1982), p. 62.

119. V. Zuckerkandl, *Sound and Symbol*, (New York: Princeton University Press, 1956), pp. 102-3.

120. Zuckerkandl relates the meaning of the organic development of a musical work directly to Goethe's notion of the "archetypal phenomenon" or "primal form." Zuckerkandl talks about the "seed" of a musical work in a way which builds upon Goethe's views on the working of the "primal form" or entelechy in the growth of organisms. He writes: "Every musical work, every finished tonal pattern, grows out of a seed that lies hidden and yet reveals itself in the pattern . . . " Zuckerkandl, *Man the Musician*, p. 171. Experienced musically, tonal relationships are a direct perception of organic necessity which is necessity in itself (not something made necessary by something outside itself). Zuckerkandl writes that "[e]very [musical] step, as it is being made, is free; once made, it is necessary" (ibid. p. 14), a statement which perfectly accords with Aristotle's views on the unity of spontaneity and necessity in organic formation (see Lear, *Aristotle: The Desire to Understand*, pp. 35-42.) The musical ear understands clearly the principle of freedom in organic form. A tone is "free" because it is the self-active wholeness of the work (its life) imparting itself moment by moment, creating itself out of itself.

121. R. Steiner, *The Boundaries of Natural Science*, (Spring Valley: Anthroposophic Press, 1983), p. 60.

122. "Other senses, whose principal function is to serve orientation on the physical stage, can attain to the perception of the phenomenon of motion in its pure essentiality [i.e., music] only under special conditions" Zuckerkandl, *Sound and Symbol*, p. 146. To this we can add that the Goethean scientific methodology represents just such "special conditions."

123. See also A. Suchantke, "The Metamorphosis of Plants as an Expression of Juvenilization in the Process of Evolution," in *Il Divano Morphologico*, No. 1, 1998, p. 56 (also published in *The Metamorphosis of Plants*, [Cape Town: Novalis Press, 1995]).

124. These shall be summarized here in the terms of another Goethean re-searcher, Margaret Colquhoun (see M. Colquhoun & A. Ewald, *New Eyes for Plants*, pp. 82-84). See also Jochen Bockemühl, "Transformation in the Foliage Leaves of Higher Plants," in *Goethe's Way of Science*, p. 116.

125. See Jochen Bockemühl, "Transformation in the Foliage Leaves of High-er Plants," in *Goethe's Way of Science*, pp. 115-128. For a discussion of the significance of this phenomenon in terms of the time-body of the plant, see A. Suchantke, "The Metamorphosis of Plants as an Expres-sion of Juvenilization in the Process of Evolution," in *The Metamorphosis of Plants*, pp. 47-68.

126. There are also musical laws which govern the ramification of branches, the variety of outward forms or "dichotomy" in venation and so forth. These have been investigated by Hans Kayser in his *Akroasis* and other works.

127. A. Husemann, *The Harmony of the Human Body*, p. 71.

128. In most plants this is true, which is why classification down to species level is most easily carried out in terms of flower characteristics. How-ever, in some plant families (and the Cruciferae is one of these) only the familial "idea" is established in the flower and one must go further, to fruit characteristics, to identify plants to the species level.

129. To say that the idea or form of the plant is in the seed, written into the genetic structure ("the organism is essentially just information"), is to confuse two entirely different things. It is like saying that the different letters that make up a sentence are the meaning of the sentence, or that the notes of a musical piece (minims, crotchets, etc., A, B, B flat, etc.) are the meaning of the piece. The genetic structure is in the first place just chemical substance, in the second place just numerical combination. The genetic structure is not a meaning or idea. The idea, or actual or-ganizational principle, is only perceived through its manifestation in the changing form of the growing plant — that is, in the plant's morphol-ogy. Of course, the genetic code has a particular function in the growth and metamorphosis of the plant. However, morphologically speaking, the idea manifests most clearly in the flower. This idea works from the future, drawing the vegetative plant toward itself.

130. Goethe speaks of the progressive refinement of substance in plants, reaching its highest stage in the corolla. He writes: "The beautiful appear-ance of the colors leads us to the notion that the material filling the petals has attained a high degree of purity." Goethe, *Scientific Studies*, p. 83.

131. In terms of the octave of the whole plant, the original seed represents the prime; the new seed, which takes form above the ground and within the fruit body, sounds the octave. In this octave of development the whole vegetative expansion carries the quality of the third, and the con-traction to the stem and receptacle sounds the consolidating force of the

fourth. The fifth to sixth are experienced in the opened out, but at the same time enclosing form of the calyx and corolla. In the form of the flower the plant most fully expresses itself or comes into its own. The hollow, enclosing space of the flower with its seed-like ovules within the ovary is in fact speaking ahead of itself, to its consummation in the new seed which falls to earth. Time, from the fifth to the octave, streams from the future; the potential of the new seed calls the floral plant toward itself. In the process of fertilization we hear the tremulous quality of the seventh, a yearning to reach and resolve into a new unity. The seed within the fruit is both a conclusion and a beginning, just as an octave is also a prime.

132. Fire names the creative or free (unconditioned or undetermined) aspect of life; the ancient Greek thinkers identified Fire with the "unmoved mover," the demiurge, the uncreated creative principle of nature which brings things into being and takes them away. The Stoic philosophers, following Aristotle, saw the "uncreated" eternal soul-warmth (Fire) of the stars (quintessence) as the origin of life, the creative "world-seeds" (*spermatagoi logoi*) through which all individualized living forms come into being. Before them Democritus (as testified by Aristotle) had spoken of the soul as "a sort of fire or hot substance" and of spherical atoms of substance as seeds of Fire. Fire, according to the Law of Opposites deriving from Pythagorean sources, is both mobility and warmth; it is formless (rarefied), an absolute homogeneity or changeless unity (polar to Earth's rigid and cold, differentiated and solid character). Fire is Zeus or the Logos, the Divine Word or Law, the rational principle within all things. The journey of knowledge was thus conceived of as toward the Sun, the fiery source of all things and all meaning. The soul, when purified, becomes like Fire; this is the ancient wisdom which found its way into Plato's Allegory of the Cave. See G.S. Kirk & J.E. Raven, *The Presocratic Philosophers*; D.E. Hahm, *The Origins of Stoic Cosmology*; D. Hyland, *The Origins of Philosophy*; and F.M. Cornford, *From Religion to Philosophy*.

133. G. Bachelard, *The Flame of a Candle*, (Dallas: The Dallas Institute, 1988), p. 45.

134. Ibid., p. 43.

135. Goethe, *Scientific Studies*, pp. 21 & 305.

136. See F. Amrine, "The Metamorphosis of the Scientist" in *Goethe's Way of Science*, p. 41.

137. Quoted in G. Adams & O. Whicher, *The Plant Between Sun and Earth*, (Boulder, CO: Shambhala, 1980), p. 3.

138. Goethe in a letter to Herder dated May 17th, 1787.

139. J. Bockemühl, *Toward a Phenomenology of the Etheric World*, p. 30.

140. The brain, if damaged, does not regenerate. By contrast, the liver (which

could be called the "heart" of the metabolic-limb organization) will regenerate even if half of it is destroyed.

141. See W. Tatarkiewicz, "Creativity; History of the Concept" in *Dialectics and Humanism* 4 (3), (1977), pp. 48-63. We find the same principle stated in Biblical tradition: the Logos, the Word, is "in the beginning," but the Word itself is uncreated, for the Word is Divine.

142. See D. Hyland, *Origins of Philosophy*, p. 153. Passages from Heraclitus give expression to the conception of Fire as rationality (measure), lawfulness and controlling power.

143. M. Heidegger makes this reference to Plato in *The Question Concerning Technology*, p. 34. For his discussion on the poetics of art, see M. Heidegger, *Early Greek Thinking*, (NY: Harper & Row, 1975), p. 19.

144. M. Heidegger, *The Question Concerning Technology*, p. 34.

145. The English poet Matthew Arnold wrote: "When [the poetic] sense is awakened in us, as to objects without us, we feel ourselves to be in contact with the essential nature of those objects, to be no longer bewildered and oppressed by them, but to have their secret, and to be in harmony with them . . . The interpretations of science do not give us this intimate sense of objects as the interpretations of poetry give us; they appeal to a limited faculty, and not to the whole man." From Matthew Arnold, *Maurice de Guérin [A Definition of Poetry]*, *Essays in Criticism, First Series* (1863), extract printed in *The Norton Anthology of English Literature* (Vol. 2), (NY: Norton & Co., 1962), p. 1424.

146. The poet Novalis, connected like Goethe with the Germanic stream of "nature philosophy," spoke of the "poetization of the sciences" whereby empirical science discovers its creative dimension, the scientist becomes the kinsman of the poet, and the sciences become a "poetically harmonious, unified work of art, a *Universalwissenschaft.*" See G.A. Smith, *The Romantic View of Science in Novalis' Notes and Fragments*, PhD Thesis, (Ann Arbor: University Microfilms, 1970), p. 77.

147. Quoted in H.B. Nisbet, *Goethe and the Scientific Tradition*, p. 67. Nisbet here discusses in general the influence of Goethe's artistic experience on his scientific work.

148. Ibid., p. 67.

149. Translation by John Barnes.

150. A fuller account of this "drama" has been told elsewhere. G. Adams and O. Whicher write: "It is the final act in the great drama when at long last the star of life — borne by the pollen-grain — descends to the waiting ovules. . . The world of the ovary is dark and moist, reminiscent of the world below the soil in which the seed once was. Above it the petals open, forming the colored hollow with the stamens within it, their anthers bearing the pollen. In the midst of these the stigma rises from below, where in the still green darkness — as though in the tomb of

Earth, yet high above it — the ovules await the messenger of the light. The pollen is at the very summit of the plant's achievement as it strives toward the light" (G. Adams & O. Whicher, *The Plant between Sun and Earth*, p. 192).

151. Goethe, *Scientific Studies*, p. 86.

152. W. Pelikan, "The Cruciferae," p. 59.

153. Ibid. p. 59. The crop and food species tend to have white or yellow flowers, and these colors are the most common for crucifers in general. However, there are ornamental species such as "honesty" (*Lunaria*) and sweet alyssum (*Alyssum* spp.) that have purple or violet floral coloration.

154. Ibid., p. 60.

CHAPTER 3

155. Quoted in D. Worster, *Nature's Economy*, (Cambridge: Cambridge University Press, 1977), p. 133.

156. G. Bachelard, *The Flame of a Candle*, p. 56.

157. T. Carlyle, "The Hero as Poet," in *On Heroes, Hero-Worship and the Heroic in History*, (London: Oxford University Press, 1957), p. 108.

158. See P. Michel, *The Cosmology of Giordano Bruno*, (London: Metheun, 1973).

159. The Platonic idea or archetype, by definition, is fixed and eternal and cannot transform. This has been one of the main obstacles to the proper understanding of how Goethe viewed the archetype. See E.-M. Kranich, *Thinking Beyond Darwin*, p. 11ff.

160. R. Steiner, *Goethean Science*, p. 19.

161. M. Riegner, "Toward a Holistic Understanding of Place: Reading a Landscape Through its Flora and Fauna," in *Dwelling, Seeing and Designing*, (NY: SUNY, 1993), p. 204.

162. It is conjectured that a fortuitous grouping of inorganic chemical substances in a particular circumstance — hypothetically a bolt of lightning in this primordial chemical soup — transformed into organic substance.

163. The evolutionary tree of the great German biologist Ernst Haeckel showed groupings of animals on the central axis; the implication was that one turned into the other, i.e., that bacteria turned into animals, lower animals into higher ones. But morphological studies have shown that this is an inaccurate portrayal. Hermann Poppelbaum explains that,

as time has passed in the development of evolutionary science, group-ings on the central axis have become side branches, each a specialized form, devoid of evolutionary potential. This includes the so-called ape-men. Imagination perceives the pathways of evolution, not as physical transformations, but as progressive manifestations of a living idea or archetype (H. Poppelbaum, *Man and Animal*, (London: Anthroposophi-cal Publishing, 1960), p. 31.

164. See A. Suchantke, *Eco-Geography*, (Great Barrington, MA: Lindisfarne Books, 2001), p. 27. James Lovelock tackles this question in his *Healing Gaia: Practical Medicine for the Planet*, (New York: Harmony Books, 1991), p. 29. Here he describes wholes such as coral reefs, beehives and the Earth itself as living organisms.

165. Suchantke's groundbreaking research will be of interest to anyone seek-ing a holistic understanding of landscapes.

166. See David L. Brierley, *In the Sea of Life Enisled: An Introduction to the teach-ing of Geography in Waldorf Education*, (Oslo: Antropos Forlag, 1998). One of Humboldt's major works was *Kosmos* (1845-62, 5 vols.); Ritter is best known for his *Erdkunde* (1822-59, 19 vols.). Brierley notes that the fa-mous French geographer Vidal de la Blanche, at the beginning of the twentieth century, also developed the idea of the Earth's surface as a "terrestrial organism."

167. See G.W.F. Hegel, *Philosophy of Nature* (1830), (trans. A. Miller), (Oxford: Clarendon Press, 1970), pp. 276-288.

168. Wherever one looks in the geological organism, whichever member of its "body" one considers, the potentially living is to be found as a kind of prophetic form. For example, the rock masses which make up the Earth's crust act as the support or basis of all processes, inorganic and organic, which take place thereon; the crust is thus implicitly a skeletal structure. The circulation system of water embraces the whole Earth, and within this hydrological system mineral substance is constantly being broken down, transported and precipitated, or raised up into plants and animals for the building of organic form. The Earth's water circulation is thus implicitly a living circulatory and digestive system. All the rhyth-mic movements of the geological organism — the tides, the seasons, the alternations of night and day — are implicitly or potentially the rhythms of life, which include the cycle of birth and death, heartbeat, peristal-sis, menstruation and ejaculation. The Sun's warmth and light, playing through the whole atmosphere and causing the movements of water, air and earth, are implicitly the vital energy of living things. Gerbert Groh-mann discusses Rudolf Steiner's notion of "prophetic form" in his book *The Plant*, Vol. I, (London: Rudolf Steiner Press, 1974), p. 201.

169. Rudolf Steiner writes: "Therefore the history of the formation of the Earth's body becomes the main thing for Goethe, and all the particulars have to fit into it. The important thing for him is the place a given rock

occupies in the totality of the Earth; the particular thing interests him only as a part of the whole. Ultimately, that mineralogical-geological system seems to him to be the correct one which recreates the processes on the Earth, which shows why precisely this had to arise at this place and that had to arise in another." R. Steiner, *Goethe's World View*, (Spring Valley, NY: Mercury Press, 1985), p. 190.

170. For a discussion of this question see the Appendix, pp. 170-172.

171. See G. Adams and O. Whicher, *The Plant between Sun and Earth*, and other books by these authors.

172. See, for example, J. Bockemühl, *Dying Forests: A Crisis in Consciousness*, (Stroud: Hawthorn Press, 1992).

173. As Hegel puts it, when a plant grows it "falls apart into a number of individuals" (*Philosophy of Nature*, p. 303).

174. Gerbert Grohmann, in *The Plant* (Vol. I), discusses Rudolf Steiner's views on this question. He writes that in the tree trunk the earth has "thrust itself up," that trunks are "protruberances of the soil." With this in mind we can liken that which grows on the trunk (i.e., the leaves) to herbaceous plants growing directly on or out of the earth. "The coming-into-leaf of a tree is the same process as the germination of a plant from seed or its sprouting from a root stock. Reversing the picture we can say: The soil is the trunk of herbaceous plants." (p. 70.)

175. There are exceptions to this — for example, the carrot and sugar beet. In these cases the color and sweetness usually associated with fruit and flower have been "carried down" into the root.

176. In some plants, such as conifers, the flowering phase is "held back" and the levity principle takes the form of volatile oils, the fiery levity nature of which is revealed when the tree burns.

177. G. Bachelard, *The Flame of a Candle*, p. 56.

178. See G. Adams & O. Whicher, *The Plant between Sun and Earth*; also O. Whicher, *Projective Geometry*, (London: Rudolf Steiner Press, 1971).

179. R. Steiner, *Goethe's World View*, p. 59.

180. G.W.F. Hegel, *The Philosophy of Nature*, p. 350.

181. G. Bachelard, *The Flame of a Candle*, pp. 43-44.

182. G.W.F. Hegel, *The Philosophy of Nature*, p. 343. The Goethean biologist Andreas Suchantke writes, in relation to flower color and form, that "[w]hat for the human being is inward remains for the plant outward, peripheral." A. Suchantke, *Metamorphosis, Evolution in Action* (Hillsdale, N.Y.: Adonis Press, 2009), p. 57.

183. Ibid., p. 356. Plants have no self-movement; they do not determine themselves over and against the world as animals do (which is the mark of a

true "self"). Plants are moved (drawn upwards) by the light, and for this reason Hegel calls the Sun the "self" of the plant (ibid., p. 306); it is not yet the sentient "self" of the animal or the self-conscious spirit of the human being.

184. H. Poppelbaum, *A New Zoology*, (Dornach: Philosophic-Anthroposophic Press, 1961), p. 165.

185. For a highly detailed discussion of the three-fold constitution of the mammals, see Wolfgang Schad, *Man and Mammals*, (N.Y.: Waldorf Press, 1977).

186. Gerbert Grohmann, *The Plant*, pp. 193-208.

187. H. Poppelbaum, *Man and Animal*; E.-M. Kranich, *Thinking Beyond Darwin*. J. Verhulst, *Developmental Dynamics in Humans and other Primates*, (Ghent, N.Y.: Adonis Press, 2003), p. 351.

188. See M. Riegner, "Horns, Hooves, Spots and Stripes: Form and Pattern in Mammals," in *Goethe's Way of Science*, pp. 177-213.

189. See H. Bortoft, *The Wholeness of Nature*, p. 108.

190. J. Bochemühl, *Awakening to Landscape*, (Dornach: Allgemeine Anthroposophische Gesellschaft, 1992), p. 268.

191. To speak of a Goethean study of place as a "creative partnership" is to distinguish it from the Deep Ecology approach which, in a spirit of radical egalitarianism, sees the human being merely as one of nature's many life forms, with no special potentiality or mission within the terrestrial life-realm. Through its analysis of the destructive tendencies of human culture and its critique of the theoretical sciences' efforts to master and control the natural world, Deep Ecology has proposed that humanity should "let things be" and interfere as little as possible with nature — for organisms are viewed as capable of realizing their potential without human interference. To Deep Ecology thinking, the very act of studying and knowing can appear as a form of interference, even destruction. See, for example, Michael Zimmerman, "Rethinking the Heidegger-Deep Ecology relationship," *Environmental Ethics* 15:3 (1993), pp. 195-224.

CHAPTER 4

192. Although it is a local name, Yabby Ponds will be capitalized in this text.

193. However, in truth, it is difficult to decide what is "natural" to Australian bush areas without considering the influence of Aboriginal culture, which occupied and molded this continent for so many millennia.

194. The wholeness of organic form, as Henri Bortoft writes in *The Wholeness of Nature*, can be seen (imaginatively, intuitively) yet cannot be depicted by a sensory representation such as a color slide (p. 384). What cannot be so depicted can, he suggests, nevertheless be embodied in works of art. Wholeness — living organization — cannot be proven, but it can nevertheless be expressed, and an artwork needs to be judged upon the basis of whether it truly or adequately succeeds in this.

195. D. H. Lawrence, *Kangaroo*, (Cambridge: University of Cambridge Press, 1994), pp. 13-15.

196. Ibid., p. 77.

197. Ibid., p. 14.

198. See D. Chapman et al., *Understanding Our Earth*, (London: Pitman,1985), p. 182.

199. E.A. Hay & A.L. McAlester, *Physical Geography*, (Englewood Cliffs, NJ: Prentice-Hall, 1984), p. 109.

200. D. Branagan et al., *An Outline of the Geology and the Geomorphology of the Sydney Basin*, (Sydney: Science Press, 1976); B. Nashar, Geology of the Sydney Basin (Brisbane: Jacaranda Press, 1967); and C. Laseron, *Ancient Australia*, (Sydney: Angus & Robertson, 1969).

201. D. Branagan, *An Outline of the Geology and the Geomorphology of the Sydney Basin*, p. 12.

202. B. Nashar, *Geology of the Sydney Basin*, p. 7.

203. N. Beadle, O. Evans & R. Carolin, *The Flora of the Sydney Region*, (Sydney: Reed, 1972), p. 9.

204. G.R. Cochrane et al., *Flowers and Plants of Victoria*, (Sydney: Reed, 1968).

205. A. Cronquist, *An Integrated System of Classification of Flowering Plants*, (NY: Colombia University Press, 1981), p. 298; and B. Morley & H. Toelken, *Flowering Plants of Australia*, (Adelaide: Rigby, 1983), p. 87.

206. N. Beadle et al., *The Flora of the Sydney Region*, p. 227.

207. A. Cronquist, *An Integrated System of Classification of Flowering Plants*, pp. 51 & 13.

208. Ibid., p. 298.

209. N. Beadle et al., *The Flora of the Sydney Region*, pp. 99-101.

210. G.H.M. Lawrence, *Taxonomy of Vascular Plants*, p. 59.

211. N. Beadle et al., *The Flora of the Sydney Region*, p. 227.

212. See the Internet site for Ku-ring-gai Chase National Park, (www.npws.

nsw.gov.au/parks/metro/Met021); go to the natural environment section for a list of animal species.

213. W. Longmore, *Honeyeaters and their Allies of Australia*, (North Ryde: Angus & Robertson, 1991), p. 172.

214. K. Simpson & N. Day, *Field Guide to the Birds of Australia*, (Ringwood: Viking Books, 1996), p. 352.

215. See C.G Sibley & J.E. Ahlquist, *Phylogeny and Classification of Birds*, (New Haven: Yale University Press, 1990).

216. K. Simpson & N. Day, *Field Guide to the Birds of Australia*, p. 9.

217. Ibid., p. 354.

218. J. Pearce, "Habitat Selection by the White-eared Honeyeater. Review of Habitat Studies" in *Emu*, Vol. 96, (1996), pp. 42-44.

219. W. Longmore, *Honeyeaters and their Allies of Australia*, pp. 170-172.

220. P.J. Higgins, *Social Organization and Territorial Behaviour of White-eared Honeyeaters at Brisbane Water National Park, New South Wales*, M. Sc. Thesis, University of Sydney, (1992), p. 40.

221. W. Longmore, *Honeyeaters and their Allies of Australia*, p. 169.

222. K. Simpson and N. Day, *Field Guide to the Birds of Australia*, p. 196.

223. P.J. Higgins, *Social Organization and Territorial Behaviour of White-eared Honeyeaters at Brisbane Water National Park, New South Wales*, pp. 50-61.

224. W. Longmore, *Honeyeaters and their Allies of Australia*, p. 172.

225. See footnote 169.

226. See the discussion in Chapter 3, in the section entitled *Evolution as Creative Process*.

227. These observations are recorded by Gaston Bachelard in his *Earth and Reveries*, pp. 300-302.

228. We discussed, in Chapter 3, how the flower is the revelation of the idea which is more or less hidden in the form of the leaf — Goethe's notion of *Steigerung* or intensification.

229. G.R. Cochrane et al., *Flowers and Plants of Victoria*.

230. G. Bachelard, *Earth and Reveries*, p. 284.

231. Ibid., p. 284.

232. Ibid., p. 284.

233. As noted in Chapter 3, Hegel called the flower the plant's "highest subjectivity."

234. Here a comparison could be made with the flora of New Zealand, a neighboring country, but relatively youthful in nature, a volcanic island formed of mainly igneous rocks. The flora is relatively unindividualized — for example, the flowers are all white, with just occasional touches of color. See the study by A. Suchantke in his *Eco-geography*, pp. 200-204.

235. See the discussion earlier in this chapter under Earth cognition: the plant organism.

236. Within the genus *Hibbertia* this loosely expansive quality is characteristic, expressed in the horizontally sprawling habit of its many herbaceous species as well as in the "vertically sprawling" habit of its climbers.

237. See Goethe's theory of color in Goethe, *Scientific Studies*, p. 155.

238. For a discussion of grouping birds based on soul type see M. Riegner, "Horns, Hooves, Spots and Stripes: Form and pattern in Mammals" in *Goethe's Way of Science*, pp. 177-212.

239. The emphasis here is on the bird's main cry and does not deal with its capacity for mimicry.

EPILOGUE

240. Quoted in R. Steiner, *The Arts and their Mission*, (Spring Valley, NY: Anthroposophic Press, 1964), p. 99.

241. Goethe, *Scientific Studies*, p. 307.

242. H. Bortoft, *The Wholeness of Nature*, p. 121.

243. Goethe, *Scientific Studies*, p. 312.

244. Ibid., p. 22.

245. See the quotation at the beginning of the Introduction.

246. Various books have been written over the last few decades which aim to show how scientific and artistic perspectives can complement each other when juxtaposed. This is specifically not what is meant by the marriage of science and art in Goethe's way of science. Examples of the complementary approach to science and art are C.H. Waddington, *Behind Appearance: A Study of the Relations between Painting and the Natural Sciences in this Century*, (Edinburgh: Edinburgh University Press, 1969), and most recently Stephen Jay Gould and Rosamond Purcell, *Crossing Over: Where Science and Art Meet*, (Three Rivers Press, 2004).

247. It is in the international Steiner (Waldorf) school movement, and related adult education and teacher training courses, that the Goethean scientific approach is currently being taught and researched.

APPENDIX

248. *Nature*, Vol. 376, July, 1995, p. 120.

249. See www.soi.stanford.edu/results/heliowhat.html and www.noao.ed/education/ighelio/solar_music.html.

250. See www.solar-center.stanford.edu/news/tornadoes.html.

251. See www.goodfelloweb.com.

252. R. Steiner, *Therapeutic Insights: Earthly and Cosmic Laws*, (Spring Valley, NY: Mercury Press, 1984), pp. 4-5.

253. See T. Schwenk's *Sensitive Chaos* for the relationship between the Sun and water vortices and G. Adams and O. Whicher, *The Plant between Sun and Earth*, for the relationship between the Sun and the form of flowers and the nature of "Sun spaces" and counter-Euclidean space.

APPENDIX

It has recently been found that the solar wind is not an explosive, random bombardment of particles through space, emanating from the Sun (as it is still described in the textbooks); rather, it is a rhythmic or periodic phenomenon. "The discovery is amazing in the light of current thinking about the solar wind, because such thinking is based on the idea that the temporal variation is predominantly the consequence of turbulence, which has a relatively smooth spectrum" (D. Gough, "Waves in the Wind").[248] This is just one example of discoveries which are challenging the conventional theories.

Another recent discovery is that the Sun displays an entirely rhythmic or harmonic character (as opposed to the explosive character of a nuclear furnace). All over the visible face of the Sun a harmonic movement is evident which has been interpreted as reflected seismic waves, in line with the conventional view that the Sun is immensely dense at its core. Yet a recent report from Stanford University, one of the principal centers of research into the nature of the Sun, states that "the Sun acts as a resonant cavity, oscillating in millions of (acoustic, gravity) modes (like a bell)."[249]

There are other major problems with the conventional view; one of these (which has been known for decades) is that the Sun's atmosphere reaches temperatures of millions of degrees centigrade whereas the surface of the Sun is only about 6,000 degrees (the heating would appear to be external rather than at the core). The Sun's core certainly appears too cool to account for light output from the surface.[250] Another problem is that the rotation rate of the Sun decreases with depth (the core hypothesis would suggest otherwise). A number of other difficulties with the "core" hypothesis are summarized and discussed by Stephen Goodfellow.[251]

Goodfellow has gathered key current evidence and presented a cogent argument for a "non-core" or "non-space" Sun. His argument centers around the hypothesis that the Sun is a plasma or super-heated gas in a magnetically unified state. This, he suggests, is a condition in which gravity is induced without a corresponding quantity of mass. The "non-space" is an absolute vacuum; Goodfellow asserts that no observable space in the universe is a vacuum because all contain some measure of mass/energy. He writes: "A hypothetical volume of absolute vacuum (non-space) will attract space;" in other words, gravity is induced without mass. The "non-core" Sun has a photospheric plasma shell and as such fits perfectly with the evidence which points to the Sun being a "resonant cavity." Goodfellow writes that a "core" Sun makes for an inefficient oscillator: "How would solar oscillations travel from a violent nuclear core, up through a somewhat inhomogeneous body and still retain precise geometrical configurations by the time they reach the photosphere?"

Fig. 52. A set of standing acoustic waves in the Sun (after Gough, "Perspectives in Helioseismology" in Science, *Vol. 272, 1996, p. 1282)*

The arguments and observations indicated above all have their starting point in the phenomenon of the Sun or phenomena associated with the Sun; a Goethean phenomenological approach to the Sun must always begin in such a way. Rudolf Steiner's approach in the realm of spiritual science goes beyond that of Goethean phenomenology, although it has a connection to it. His views are remarkable in that they appear to anticipate the discoveries which science is now making in relationship to the Sun. Some of Steiner's ideas have been discussed by G. Blattmann in his *The Sun: Ancient Mysteries and a New Physics*, (Great Barrington, MA: Anthroposophic Press, 1996). Certainly a great deal has yet to be said on this topic and Blattmann's presentation is initial. He suggests (following Steiner) that the Sun is a hollow space which is the center of a vortex.

Now, it is just this idea of a solar vortex which was uppermost in the mind of Descartes, and which Newton rejected in favor of a model inspired by magnetism — that is, the attraction of massive bodies through empty space. What Newton did not have was the understanding, through modern projective geometry, of the action of universal levity forces. According to this understanding, the space of the Sun is not a three-dimensional space at all but a counter-Euclidean space, a "Sun space." This "Sun space" is governed, not by an infinitude without (the plane at infinity in terms of three-dimensional or Euclidean space), but by an infinitude within, a center of infinite receptivity which occurs in non-Euclidean or ethereal space. According to Steiner, the Sun's ethereal center has both "negative materiality" and an infinitude of levity force. Steiner indicates:

> You can think of [the space of the Sun] in [this] way: it is not only empty but you could say that it exerts suction because there is less than 0 in it . . . the Sun therefore has an inward suction; it does not exert pressure like a gas. The Sun space is filled with negative materiality.[252]

We can see an image of this space in the hollow space of the vortex in water; we can see another in the hollow chalice of the flower.[253]

About the Author

Nigel Hoffmann was born in Sydney, Australia, in 1953. He studied fine arts and botany at the University of Sydney and attained a doctorate from the University of Newcastle, in the area of Goethean science. In 1986 he founded the magazine *Transforming Art*, which he edited for ten years; one of the main themes explored in this publication was the relationship of art to science. He has taught a number of short courses in Goethean science and, in recent years, has been teaching at Waldorf schools in Melbourne, Basel (Switzerland) and in Sydney. He is the author of *The University at the Threshold: Orientation through Goethean Science*, Rudolf Steiner Press, 2020.